INTERNATIONAL CENTRE FOR MECHANICAL SCIENCES

COURSES AND LECTURES - No. 57

PETER SAGIROW
STUTTGART UNIVERSITY

STOCHASTIC METHODS IN THE DYNAMICS OF SATELLITES

COURSE HELD AT THE DEPARTMENT
FOR GENERAL MECHANICS
OCTOBER 1970

UDINE 1970

SPRINGER-VERLAG WIEN GMBH

ISBN 978-3-211-81092-7 ISBN 978-3-7091-2870-1 (eBook)
DOI 10.1007/978-3-7091-2870-1

P R E F A C E

This textbook is based partly on the author's lectures on Satellite Dynamics at the University of Stuttgart and partly on an advanced seminar on Modern Stochastic Methods held by Mr. L. Arnold and the author at Stuttgart during three semesters in 1969 and 1970. The author is deeply grateful to the participants of the seminar for discussions, helpful remarks, and the reading of the manuscript.

The author is very much indebted to Professor Sobrero, CISM, for the invitation to deliver this lecture at Udine and for understanding cooperation.

Stuttgart, 14-9-1970.

1. Introduction

The librational motion of a satellite depends on the gravity gradient, aerodynamic, and magnetic torques, on the moments of inertia of the satellite and on some less significant influences as solar radiation, the electrical field of the earth, the meteorite impacts. Only the gravity gradient torque can be considered with a great degree of accuracy as deterministic. Aerodynamic and magnetic torques can be treated as deterministic quantities only in a very rough first approximation. In reality these torques always have stochastic components caused by the fluctuations of the atmospheric density and the earth's magnetic field. Furthermore, the calculation of these torques for a given satellite is based on assumptions whose nature is indeed stochastic. The same is true for influences caused by solar radiation, meteorite impacts, and electrical field of the earth. The moments of inertia of a satellite are deterministic only for completely rigid bodies. The uncontrolled thermo-elastic oscillations of the stabilizing rods, antennae, and sun cell panels, the motion of the crew, the wobbling of liquid in the tanks etc. make the concept of an ideally rigid satellite questionable. Thus, the moments of inertia have to be regarded as stochastic quantities too. In consequence of the stochastical nature of the mentioned outer and inner influences the librational motion of a satellite is a stochastic process and has to be described by stochastic differential equations. The first conception of these equations was given in the late forties by Itô [1]. In the following twenty years this first conception was developed by the works of Skorokhod, Bucy, Kushner, Khasminski a.o. The first books on this subject appeared in the last few years : Bucy [2], Kushner [3], Gikhmanand Skorokhod [4], Khasminski [5]. Today the Theory of Stochastic Differential Equations and the Theory of Stochastic Stability are powerful tools in the applications and especially in the dynamics of satellites.

In the present contribution the stochastic models of librating satellites are considered. As the stochastic methods are new, a brief survey on stochastic processes, stochastic differential equations and stochastic stability is given first. Then some less complicated problems of stochastic dynamics of satellites are discussed. The stochastic stability of the considered systems is investigated as well by Liapunov techniques, as in first approximation, and in the mean square. With the exception of the last example in section 6.2 all problems are solved completely. Some known results on stochastic stability of satellites are improved and new results are obtained. In the last section 6.2 a satellite with variable moments of inertia is considered. This example shows the feasibilities of stochastic modeling and the difficulties which can arise. This example is discussed only.

2. Stochastic Dynamical Systems

The correct way from the intuitive idea of probability to the theory of stochastic stability is long and fatiguing : Measure Theory, Probability Theory, Theory of Stochastic Processes, Theory of Stochastic Differential Equations and finally Stochastic Stability. This tour is seldom done completely by mathematicians and is unacceptable for a man in applications. On the other hand, just the results at the end of this cumbersome way - the criteria of stochastic stability - are of great interest in mechanics and engineering. Therefore, here the attempt will be made to reach the correct results on stochastic stability by a non rigorous but brief treatment based more on faith and intuition than on an extensive mathematical background and sophisticated proofs. Only some elementary knowledge on probability is supposed. Section 2.1 contains notations and basic definitions on density functions, moments and stochastic processes. In 2.2 dynamic systems corrupted by noise are introduced. Sections 2.3 to 2.6 give a hint of an outline of the theory of stochastic differential equations.

Finally in 2.7 the basic definitions and results on stochastic stability are collected. As the topics of 2.3 to 2.7 are relatively new and the few books treating them are written on a very high level, the formulae and theorems most important for the applications in sections 3 to 6 are deduced by heuristic considerations. For strong proofs see [3], [4], [5], [6] and consult previously [8], [9], [10] - if necessary.

2.1 Basic Definitions

2.1.1 Probability Density Functions

With a scalar random quantity x is associated the probability density function $f(x)$ defined by the property

$$P(x_1 \leq x < x_2) = \int_{x_1}^{x_2} f(x)dx ,$$

where $P(x_1 \leq x < x_2)$ denotes the probability of the event $x_1 \leq x < x_2$.

For normally or Gaussian distributed random quantities the density is given by

$$f(x) = \frac{1}{\sqrt{2\pi}\,\sigma} \exp\left[-\frac{(x-\bar{x})^2}{2\sigma^2}\right] \qquad (2.1)$$

where $\bar{x}$ and σ^2 are the mean value and the variance of x, respectively (see 2.1.2).

2.1.2 Moments

Random quantities can be described by means of their moments. The moments are defined by

$$Ex^k = \int_{-\infty}^{\infty} x^k f(x)dx , \quad k = 1,2,3,\ldots$$

and by

$$E(x - Ex)^k = \int_{-\infty}^{\infty} (x - Ex)^k f(x)dx , \quad k = 1,2,3,\ldots$$

The most important moments are the mean or average value or expected value

$$\bar{x} = Ex \tag{2.2}$$

and the variance

$$\sigma^2 = E(x - \bar{x})^2 . \tag{2.3}$$

For normally distributed random quantities all higher moments $(K > 2)$ can be expressed by the first moment $\bar{x}$ and the second moment σ^2.
The quantity σ is the standard deviation.

2.1.3 Random Vectors

The first and the second moments of a

random vector

$$x = \begin{bmatrix} x_1 \\ x_2 \\ \vdots \\ x_n \end{bmatrix}$$

are defined by

$$\bar{x} = Ex = \begin{bmatrix} Ex_1 \\ Ex_2 \\ \vdots \\ Ex_n \end{bmatrix} \tag{2.4}$$

and by the covariance matrix

$$P = E(x - \bar{x})(x - \bar{x})^* = \tag{2.5}$$

$$= \begin{bmatrix} E(x_1 - \bar{x}_1)(x_1 - \bar{x}_1) & E(x_1 - \bar{x}_1)(x_2 - \bar{x}_2) \ldots & \\ E(x_2 - \bar{x}_2)(x_1 - \bar{x}_1) & E(x_2 - \bar{x}_2)(x_2 - \bar{x}_2) \ldots & \\ \cdots & \cdots & \cdots \\ E(x_n - \bar{x}_n)(x_1 - \bar{x}_1) & E(x_n - \bar{x}_n)(x_2 - \bar{x}_2) \ldots & E(x_n - \bar{x}_n)(x_n - \bar{x}_n) \end{bmatrix} .$$

The diagonal elements of P are the variances and the off-diagonal elements the covariances of the vector components. The standard square deviation is given by the trace of P:

$$\operatorname{tr} P = E\Big(\sum_{i=1}^{n} (x_i - \bar{x}_i)^2\Big) . \tag{2.6}$$

The covariance matrix is symmetric and positive definite.

The density function of a normally distributed n-vector is :

$$(2.7) \quad f(x_1,x_2,..,x_n) = \frac{1}{\sqrt{(2\pi)^n}\sqrt{\det P}} \exp\left[-\frac{1}{2}(x-\bar{x})^* P^{-1}(x-\bar{x})\right] .$$

2. 1. 4 Stochastic Processes

A random function of time is called a stochastic process. The mean, the variance and the covariance matrix of a scalar process $x(t)$ are

$$\bar{x}(t) = Ex(t)$$

$$\sigma^2(t) = \text{var}\, x(t) = E(x(t) - \bar{x}(t))^2$$

$$\text{cov}\, x(t)x(s) = R(t,s) = E(x(t) - \bar{x}(t))(x(s) - \bar{x}(s))$$

where R is the well known correlation function. Only two of these three quantities are independent, for

$$(2.8) \qquad \sigma^2(t) = R(t,t) .$$

For stationary processes mean and variance are constants

$$\bar{x} = \text{const}, \qquad \sigma^2 = \text{const},$$

whereas R depends only on $\tau = |t - s|$. In consequence for stationary processes

$$\sigma^2 = R(0) . \tag{2.9}$$

2. 1. 5 Markov Processes

A stochastic process $x(t)$ is a Markov process if for every t_0 the probability of the future state $x(t)$, $t > t_0$, depends only on the present state $x(t_0)$ and is independent of the past, i.e. is independent of the process history $x(t)$, $t < t_0$.
Markov processes are the stochastic analogues to deterministic processes without time-lag.

2. 1. 6 Process with Independent Increments

If for arbitrary times

$$0 = t_0 < t_1 < t_2 < \ldots < t_k$$

the corresponding process increments

$$x(t_1), x(t_2) - x(t_1), \ldots, x(t_k) - x(t_{k-1})$$

are independent, the process is called a process with independent increments and it is

$$x(t_k) = \sum_{i=1}^{k} \left(x(t_i) - x(t_{i-1})\right) .$$

2.1.7 Wiener Processes

A normally distributed process $w(t)$ with independent increments is a Wiener process if

1 $w(0) = 0$

2 $Ew(t) = 0$

3 $E[w(t+h) - w(t)]^2 = a^2 h, \quad h > 0, \; a^2 = \text{const}.$

or 3' $E[w(t+h)w(t)] = a^2 t.$

If $w(t)$ is a Wiener process every process $cw(t)$, $c =$ $=$ const, is a Wiener process too. Thus, every Wiener process can be expressed by the standard Wiener process with $Ew^2(t) = t$.
The Wiener process is continuous with probability 1 and is nowhere differentiable.

2.1.8 White Noise

Given a stochastic process x the quantities $x(t)$ and $x(s)$ are connected by the correlation function $R(t,s)$ or $R(\tau)$, $\tau = |s-t|$. For many applications it is convenient to introduce the fiction of a "completely stochastic" process. For this process- called the white noise- the correlation function is supposed to be the Dirac delta-function :

$$R(\tau) = \delta(\tau) \tag{2.10}$$

i.e. the process values $x(t)$ and $x(s)$ are supposed to be uncorrelated for $t \neq s$. In consequence of (2.10) the spectral density of the white noise is a constant. Thus, the white noise involves all frequencies uniformly and has an infinite energy !

No real process has the properties mentioned above. In spite of this fact the white noise is a useful fiction in stochastic dynamics. Using this fiction as an initial concept a great number of real stochastic processes can be constructed :

1. The Wiener process $w(t)$ is the integral of the white noise $\xi(t)$

$$w(t) = \int_0^t \xi(s)ds \tag{2.11}$$

or

$$dw(t) = \xi(t)dt\,. \tag{2.12}$$

<u>Caution</u> : Instead of the correct relations (2.11) and (2.12) the brief but incorrect (see sec. 2.1.7) notation

$$\xi(t) = \dot{w}(t) \tag{2.13}$$

for the white noise is widely used in applications as in sec. 3 to 6. In any case this shorthand notation has to be understood in the sense of (2.11) or (2.12)!

2. Using the white noise $\xi(t)$ as the input of a linear system colored noise is obtained. For instance, the solution (or output) of the linear system

$$(2.14) \qquad \dot{\eta}(t) = -a\eta(t) + b\xi(t), \qquad a > 0$$

is a colored noise with the correlation function

$$(2.15) \qquad R(\tau) = \frac{b}{2a} e^{-a|\tau|} .$$

Linear systems used for the generation of colored noise are called shaping filters.

2.1.9 Diffusion Processes

Roughly speaking a diffusion process is a process whose velocity is composed of a deterministic part (the drift) and of a white noise (the diffusion). Diffusion processes are continuous Markov processes.

2.2 Stochastic Dynamical Systems

Let the state $x(t)$ of a deterministic system satisfy the equation

$$x(t + \Delta) = x(t) + f(x(t))\Delta .$$

If the system is corrupted by the noise $v(t)$ the state becomes $x(t) - \sigma v(t)$ and the system equation will be

$$x(t + \Delta) - \sigma v(t + \Delta) = x(t) - \sigma v(t) + f(x(t) - \sigma v(t))\Delta$$

where σ denotes some proportionality factor. Expanding the function f and supposing $\sigma v(t)$ and Δ to be small we get

$$x(t + \Delta) - \sigma v(t + \Delta) = x(t) - \sigma v(t) + f(x(t))\Delta$$

or

$$x(t + \Delta) - x(t) = f(x)\Delta + \sigma(v(t + \Delta) - v(t)).$$

If no special information on the nature of the disturbance $v(t)$ is available following assumptions on $v(t)$ seem to be natural :

1. there is no noise at the start, i.e. $v(0) = 0$,
2. there is no systematic error, i.e. $Ev(t) = 0$,
3. The increments of $v(t)$ are independent and normally distributed,
4. The dispersion increases with time , i.e. $Ev^2(t) = a^2 t$ where a^2 is some coefficient.

Thus, it is natural to assume $v(t)$ to be a Wiener process (see 2.1.7). Then the system equation becomes

$$x(t + \Delta) - x(t) = f(x)\Delta + \sigma(w(t + \Delta) - w(t)) . \quad (2.16)$$

At this point it is tempting to divide by Δ and to replace the difference equation by

$$\dot{x} = f(x) + \sigma\dot{w}. \tag{2.17}$$

This way was customary until the early sixties and leads to correct results of the correlation theory if the proportionality factor σ is independent of the state variables.

However, in many practical problems (see sec. 3.5.6) the factor σ depends explicitly on the state : $\sigma = \sigma(x)$ or $\sigma = \sigma(x,t)$. In this case eq. (2.16) has to be replaced by the so-called stochastic differential equation

$$dx = f(x)dt + \sigma(x)dw(t) \tag{2.18}$$

or

$$dx = f(x,t)dt + \sigma(x,t)dw(t). \tag{2.18'}$$

A correct interpretation of the solution of eqs. (2.18) and (2.18') turns out to be possible by means of the corresponding integral equations

$$x(t) = x(t_0) + \int_{t_0}^{t} f(x(\tau))d\tau + \int_{t_0}^{t} \sigma(x(\tau))dw(\tau) \tag{2.19}$$

or

$$x(t) = x(t_0) + \int_{t_0}^{t} f(x(\tau),\tau)d\tau + \int_{t_0}^{t} \sigma(x(\tau),\tau)dw(\tau) \,. \quad (2.19')$$

The first integrals in eqs. (2.19) and (2.19') are usual, the second integrals - the so-called "stochastic integrals" - are yet to be defined. The purpose of the rest of this section is to give the reader an idea of stochastic differential equations (2.18), (2.19) and to obtain some stability theorems which are important for the applications.

2.3 Stochastic Integrals

As stochastic integrals the following integrals

$$\int_0^t \sigma(\tau)\,dw(\tau) \,, \quad (2.20)$$

$$\int_0^t \sigma(\tau,w(\tau))\,dw(\tau) \,, \quad (2.21)$$

$$\int_0^t \sigma(\tau,x(\tau))\,dw(\tau) \quad (2.22)$$

are denoted. Here, $w(t)$ and $x(t)$ are a Wiener and a Markov process, respectively.

If $\sigma(t)$ is continuous and

$$\int_0^t \sigma^2(\tau)d\tau < \infty \,,$$

the integral (2.20) is defined by

$$\int_0^t \sigma(\tau)dw(\tau) = \lim_{\substack{n\to\infty \\ \lambda\to 0}} \sum_{k=0}^{n-1} \sigma(t_k)[w(t_{k+1}) - w(t_k)]$$

where

$$0 = t_0 < t_1 < t_2 < \ldots < t_n = t, \quad \lambda = \max_k(t_{k+1} - t_k).$$

According to this definition and regarding the properties of a Wiener process (see 2.1.7) we obtain the basic properties of the stochastic integral (2.20) :

$$\int_0^t (a\sigma_1 + b\sigma_2)dw = a\int_0^t \sigma_1 dw + b\int_0^t \sigma_2 dw \tag{2.23}$$

$$E\int_0^t \sigma dw = 0 \tag{2.24}$$

$$E\left(\int_0^t \sigma\, dw\right)^2 = \int_0^t E\sigma^2 d\tau. \tag{2.25}$$

The definition of the integrals (2.21) and (2.22) is more complicated but similar in principle. The properties (2.23) to (2.25) hold for the integrals (2.21), (2.22) too.

Remark : The strong definition of stochastic integrals was first given by Itô [1] as above. A different definition is due to Stratonovich [7]. In this contribution only Itô integrals are used.

2.4 Stochastic Differentials

A process $x(t)$ satisfying for every $0 \leq t_1 < t_2 \leq T$ the relation

$$x(t_2) - x(t_1) = \int_{t_1}^{t_2} f(t)dt + \int_{t_1}^{t_2} \sigma(t)dw(t) \qquad (2.26)$$

with

$$\int_0^t |f(t)|dt < \infty , \quad \int_0^t \sigma^2(t)dt < \infty \qquad (2.27)$$

is said to possess the differential

$$dx = f(t)dt + \sigma(t)dw(t) . \qquad (2.28)$$

2.5 Ito Formulae

In consequence of the properties of the Wiener process $w(t)$ and of the stochastic integral the calculus for the stochastic differentials differs from the well known calculus for deterministic differentials.

2.5.1 dg(w)

Given a scalar Wiener process $w(t)$ and a scalar differentiable function g, what is $dg[w(t)]$?

Consider

$$g(w + dw) = g(w) + g'(w)dw + \frac{1}{2}g''(w)(dw)^2 + \ldots$$

$$dg(w) = g(w + dw) - g(w) = g'(w)dw + \frac{1}{2}g''(w)(dw)^2 + \ldots$$

If w is deterministic the term $(dw)^2$ is neglected as a term of higher order. In the stochastic case

$$dw(t) = w(t + dt) - w(t)$$

and

$$(dw)^2 = \left[w(t + dt) - w(t)\right]^2 .$$

According to the property 3 of the Wiener process

$$E\left[dw(t)\right]^2 = E\left[w(t + dt) - w(t)\right]^2 = dt$$

the term $[dw]^2$ is of first order at least in the mean and can't be neglected. Thus, the formula

$$dg(w) = g'(w)dw + \frac{1}{2}g''(w)dt \tag{2.29}$$

seems to hold in the stochastic case.

The formula (2.29) can be proofed rigorously by means of the strong theory of stochastic integrals and has interesting consequences on the integration of stochastic functions.

For instance :

$$dw^2(t) = 2w(t)dw + dt$$

$$w^2 = 2\int_0^t w\,dw + t$$

$$\int_0^t w\,dw = \frac{1}{2}w^2 - \frac{1}{2}t\,. \qquad (2.30)$$

2.5.2 dg(t,x(t))

Given a scalar stochastic process $x(t)$ with the differential

$$dx = f\,dt + \sigma\,dw$$

what is $dg(t,x(t))$? Consider

$$dg(t,x(t)) = g(t+dt, x+dx) - g(t,x) =$$

$$= g_t\,dt + g_x\,dx + \frac{1}{2!}g_{tt}(dt)^2 + \frac{1}{2!}g_{tx}\,dt\,dx +$$

$$+ \frac{1}{2!}g_{xx}(dx)^2 + \ldots =$$

$$= g_t\,dt + g_x(f\,dt + \sigma\,dw) + \frac{1}{2}g_{tt}(dt)^2 +$$

$$+ \frac{1}{2}g_{tx}(f\,dt + \sigma\,dw)dt + \frac{1}{2}g_{xx}(f\,dt + \sigma\,dw)^2 + \ldots$$

Neglecting the higher order terms as $(dt)^2$ and dtdw and replacing $(dw)^2$ by (dt) we obtain finally the famous Itô formula

(2.31) $$dg(t,x) = (g_t + fg_x + \frac{1}{2}\sigma^2 g_{xx})dt + \sigma g_x dw .$$

Remark : Regarding the fact that the increment dw is independent of x and thus

$$E\sigma g_x dw = E\sigma g_x E dw = 0$$

we get from relation (2.31) the mean of the derivative dg/dt :

(2.32) $$E\frac{dg(t,x)}{dt} = E\left[g_t + fg_x + \frac{1}{2}\sigma^2 g_{xx}\right].$$

The operator

(2.33) $$L = \frac{\partial}{\partial t} + f\frac{\partial}{\partial x} + \frac{1}{2}\sigma^2\frac{\partial^2}{\partial x^2}$$

corresponding to the right side of (2.32) is widely used in the stochastic Liapunov theory (see 2.7).

2.5.3 $dx_1 x_2$

Given two processes $x_1(t)$ and $x_2(t)$ with the differentials

$$dx_1 = f_1 dt + \sigma_1 dw$$

$$dx_2 = f_2 dt + \sigma_2 dw ,$$

what is $dx_1 x_2 = ?$

Consider again

$$dx_1x_2 = x_1(t + dt)x_2(t + dt) - x_1(t)x_2(t) =$$

$$= (x_1 + dx_1)(x_2 + dx_2) - x_1x_2 =$$

$$= x_1dx_2 + x_2dx_1 + dx_1dx_2 =$$

$$= x_1dx_2 + x_2dx_1 + f_1f_2(dt)^2 + (f_2\sigma_1 + f_1\sigma_2)dt\,dw +$$

$$+ \sigma_1\sigma_2(dw)^2 .$$

And thus

$$dx_1x_2 = x_1dx_2 + x_2dx_1 + \sigma_1\sigma_2dt . \qquad (2.34)$$

If x_1 and x_2 are vector processes with the differentials

$$dx_1 = f_1dt + \sigma_1dw \qquad f_k = \begin{bmatrix} f_{k1} \\ \vdots \\ f_{kn} \end{bmatrix}, \quad \sigma_k = \begin{bmatrix} \sigma_{k1} \\ \vdots \\ \sigma_{kn} \end{bmatrix}, \quad k = 1, 2$$

$$dx_2 = f_2dt + \sigma_2dw$$

the analogon of (2.34) is

$$dx_1x_2^* = x_1dx_2^* + (dx_1)x_2^* + \sigma_1\sigma_2^*dt . \qquad (2.35)$$

2.6 Stochastic Differential Equations

2.6.1 Definitions and Notations

Generalizing the concept of the stochastic differential we introduce the stochastic differential equation

$$(2.36) \qquad dx(t) = f(t,x(t))dt + \sigma(t,x(t))dw(t)$$

with initial condition $x(0)$ as a brief notation for the integral equation

$$(2.37) \quad x(t) = x(0) + \int_0^t f(s,x(s))ds + \int_0^t \sigma(s,x(s))dw(s) .$$

The solution $x(t)$ of eq. (2.37) is supposed to possess the differential

$$dx(t) = \bar{f}(t)dt + \bar{\sigma}(t)dw(t)$$

with

$$\bar{f}(t) = f(t,x(t)) , \quad \bar{\sigma}(t) = \sigma(t,x(t))$$

and

$$\int_0^t \bar{f}(s)ds < \infty , \quad \int_0^t \bar{\sigma}^2(s)ds < \infty .$$

In the older literature and in applications eq. (2.36) often is written as

$$\dot{x} = f(t,x) + \sigma(t,x)\xi$$

where ξ stands for the white noise.

Eq. (2.36) can be interpreted as a scalar equation and as a vector equation. In the last case, x and f are n-vectors, w is a r-vector with r Wiener processes as components, and σ is a $n \times r$ matrix.

In the linear case the scalar eq. (2.36) reads as

$$dx = (\alpha + \beta x)dt + (\gamma + \delta x)dw , \tag{2.38}$$

where α , β , γ , δ are constant or time-dependent scalars. A vector eq. (2.36) has to be replaced in the linear case by

$$dx = (a + Ax)dt + \sum_{i=1}^{r}(b_i + B_i x)dw_i , \tag{2.39}$$

where a and b_i are vectors, whereas A and B_i denote matrices.

2. 6. 2 General Properties

For a fixed initial condition $x(0)$ a unique and continuous solution of eq. (2.36) exists with probability 1 if the conditions

$$|f(t,x) - f(t,y)| + |\sigma(t,x) - \sigma(t,y)| \leq K|x - y|$$

$$f^2(t,x) + \sigma^2(t,x) \leq K^2(1 + x^2) \tag{2.40}$$

$$K = const$$

are satisfied.

If eq. (2.36) possesses a unique solution this solution

is a Markow process.

In consequence of 2.1.9 it follows : if conditions (2.40) are satisfied, the solution of eq. (2.36) is a diffusion process. Under some additional assumptions it can be shown : every diffusion process $x(t)$ can be considered as the solution of a stochastic differential equation (2.36). Thus, solutions of stochastic differential equations and diffusion processes correspond mutually.

2. 6. 3 Explicit Solutions of Linear Equations

Consider first the scalar linear homogeneous equation

$$(2.41) \qquad dx = \beta x\,dt + \delta x\,dw \;, \quad \beta, \delta \text{ - constants} .$$

Suppose $x(0) > 0$. Then, for sufficient small t the new variable y can be introduced as

$$y(t) = \ln x(t) .$$

By the Itô formula (2.31) follows

$$\begin{aligned} dy &= (y_t + f y_x + \frac{1}{2}\sigma^2 y_{xx})dt + \sigma y_x\,dw = \\ &= (0 + \beta x \frac{1}{x} + \frac{1}{2}\delta^2 x^2 \cdot - \frac{1}{x^2})dt + \delta x \frac{1}{x} dw = \\ &= (\beta - \frac{1}{2}\delta^2)dt + \delta\,dw(t) . \end{aligned}$$

By the definition of stochastic differentials (see 2.4) we have

$$y(t) = y(0) + \int_0^t (\beta - \frac{1}{2}\delta^2)ds + \int_0^t \delta\, dw(s),$$

$$y(t) = y(0) + (\beta - \frac{1}{2}\delta^2)t + \delta w(t).$$

Now, regarding

$$x(t) = e^{y(t)}$$

the solution of eq. (2.41) is obtained as

$$x(t) = x(0)e^{(\beta - \frac{1}{2}\delta^2)t + \delta w(t)}. \tag{2.42}$$

This solution holds as long as $x(t) > 0$.

As $x(0)$ was supposed to be positive, the solution holds obviously for all $t \geq 0$. If $x(0) < 0$ the variable y has to be defined by $y(t) = \ln[-x(t)]$ and then the same solution (2.42) is obtained. Thus, by (2.42) the general solution of eq. (2.41) is given.

In a similar way eq. (2.41) can be solved if $\beta = \beta(t)$ and $\delta = \delta(t)$. For the solution of the inhomogeneous eq. (2.38) see [4].

The explicit solution of linear systems (2.39) seems not to be possible in general. Only for the case $B_i \equiv 0$ the system (2.39) can be solved explicitly. Suppose $a \equiv 0$ and write (2.39) as

$$dx = Axdt + Bdw \tag{2.43}$$

with A, B matrices and w a Wiener vector process. The solution of (2.43) is

$$x(t) = \Phi(t,t_0)x(t_0) + \int_{t_0}^{t}\Phi(t,s)Bdw(s), \tag{2.44}$$

where Φ denotes the fundamental matrix of A. If $A = \text{const}$, Φ is given by

$$\Phi(t,t_0) = e^{A(t-t_0)}. \tag{2.45}$$

If A depends on time the fundamental matrix can generally not be given in closed form. For the calculation of Φ see for instance [24].

2.7 Stochastic Stability [3],[5]

As an explicit solution of stochastic differential equations can be obtained only in the simplest cases, stability investigations of stochastic systems are of high importance. In the present textbook only few basic facts on stochastic stability can be given. Nevertheless, these few factswill be sufficient to discuss in the following sections 3 to

6 various simpler stability problems of satellite dynamics. Extensive stability studies in theory and applications require a solid background in probability theory, see for instance [8], [9], [10].

2.7.1 Stability in the Mean

Consider the linear stochastic differential equation

$$dx = A(t)x\,dt + \sum_{i=1}^{r} B_i(t)x\,dw_i(t). \tag{2.46}$$

Applying the expectation operator E to both sides of eq. (2.46)

$$dEx = A(t)Ex\,dt + \sum_{i=1}^{r} B_i(t)E(x\,dw_i)$$

and accounting the independence of $x(t)$ and $dw_i(t)$

$$E(x\,dw_i) = ExEdw_i = ExE(w_i(t+dt) - w_i(t)) = 0,$$

we obtain

$$d\bar{x} = A(t)\bar{x}\,dt$$

or

$$\frac{d\bar{x}}{dt} = A(t)\bar{x}, \tag{2.47}$$

where $\bar{x} = Ex$.

Thus, the mean process $\bar{x}(t)$ satisfies the undisturbed deterministic eq. (2.47).

The equilibrium position $x = 0$ of eq. (2.46) is said to be (asymptotically) stable in the mean, if the equilibrium position $\bar{x} = 0$ of eq. (2.47) is (asymptotically) stable.

For the check of the stability in the mean it is sufficient to check the stability of the undisturbed eq. (2.47).

Remark : If the given stochastic system is of the form

$$dx = A(t)x\,dt + B(t)dw \tag{2.48}$$

the mean process $\bar{x}(t)$ satisfies the same eq. (2.47). However, here it is not possible to speak of the stability in the mean of $x = 0$. Here, the deterministic stability of $\bar{x} = 0$ with respect to the disturbance $B(t)dw$ can be discussed only (see remark 1 in 2.7.2).

2.7.2 Stability in the Mean Square

Consider the linear stochastic differential vector equation

$$dx(t) = A(t)x(t)dt + B(t)x(t)dw(t) \tag{2.49}$$

with A and B square matrices and $w = w(t)$ a scalar Wiener process.

According to Itô formula (2.35) from

$$dx = Axdt + Bxdw$$

$$dx^* = x^*A^*dt + x^*B^*dw$$

follows

$$dxx^* = xdx^* + (dx)x^* + Bxx^*B^*dt =$$

$$= x(x^*A^*dt + x^*B^*dw) + (Axdt + Bxdw)x^* +$$

$$+ Bxx^*B^*dt =$$

$$= xx^*A^*dt + xx^*B^*dw + Axx^*dt + Bxx^*dw +$$

$$+ Bxx^*B^*dt .$$

Taking the mean value E of both sides of this equation we obtain

$$\begin{aligned} dExx^* &= Exx^*A^*dt + Exx^*B^*dw + AExx^*dt + \\ &\quad + BExx^*dw + BExx^*B^*dt . \end{aligned} \tag{2.50}$$

As $x(t)$ and the increment $dw(t)$ are independent and w is scalar with $Ew = 0$, we have

$$Exx^*B^*dw = E(xx^*dw)B^* = Exx^*EdwB^* = 0$$

$$BExx^*dw = BE(xx^*)Edw = 0 .$$

Using for the covariance matrix the notation

$$P = Exx^*$$

eq. (2.50) can now be rewritten as

$$dP = (AP + PA^* + BPB^*)dt$$

or as

$$\frac{dP}{dt} = A(t)P + PA^*(t) + B(t)PB^*(t) \ . \tag{2.51}$$

Thus, the covariance matrix of the solution of a linear stochastic differential equation satisfies a deterministic matrix differential equation.

As the matrix

$$P = \begin{bmatrix} p_{11} & p_{12} & \cdots & p_{1n} \\ p_{21} & p_{22} & \cdots & p_{2n} \\ \vdots & \vdots & & \\ p_{n1} & p_{n2} & & p_{nn} \end{bmatrix}$$

is symmetric

$$p_{ij} = Ex_i x_j = Ex_j x_i = p_{ji}$$

there are $\frac{1}{2}n(n + 1)$ different covariances p_{ij} and eq. (2.51) is equivalent to a linear system of order $\frac{1}{2}n(n + 1)$. Denoting the vector of the different (arbitrarily ordered) covariances p_{ij} by p we can write

(2.51) as

$$\dot{p} = M(t)p . \quad (2.52)$$

Thus, to check the stability of the second moments it is sufficient to check the stability of a deterministic linear differential equation (2.52) of order $\frac{1}{2}n(n + 1)$. Especially, if A and B are constant, M is constant too and the well-known stability criteria of Routh, Hurwitz etc. can be applied, [16].

The equilibrium position $x = 0$ of eq. (2.49) is said to be (asymptotically) stable in the mean square if the equilibrium position $P = 0$ of the covariance eq. (2.51) (or $p = 0$ of eq. (2.52)) is (asymptotically) stable.

Often the stability in mean square is called the stability of second moments. If the given dynamical system is influenced by a Wiener vector process $w(t)$ with independent components $w_i(t)$

$$dx = A(t)x\,dt + \sum_{i=1}^{r} B_i(t)x\,dw_i(t) \quad (2.53)$$

the covariance matrix P satisfies the equation

$$\frac{dP}{dt} = A(t)P + PA^*(t) + \sum_{i=1}^{r} B_i(t)PB_i^*(t) . \quad (2.54)$$

Remark 1 : Given a system

$$dx = A(t)x\,dt + B(t)dw \tag{2.55}$$

where w is a Wiener process with independent components, the covariance matrix satisfies

$$\frac{dP}{dt} = A(t)P + PA^*(t) + B(t)B^*(t)\,. \tag{2.56}$$

Speaking of stability in this case the deterministic stability of the equilibrium position $\bar{x} = 0$ of eq. (2.47) with respect to disturbance Bdw is meant (see remark 2.7.1).

Remark 2 : For large n it is rather cumbersome to check the stability in mean square via covariance matrix. If the system is described by one stochastic differential equation of higher order it is more convenient to use special criteria given in [5]. However, the reduction of a system of first order stochastic differential equations to one equation of higher order is often impossible or at least questionable.

2.7.3 Two Theorems on the Stability of the Moments

The stability of the covariance matrix implies the stability of the variances Ex_i^2 and thus the stability of the means Ex_i:

Theorem 2.1 : The stability in mean square implies the stability in the mean (or : the stability of the second moments implies the stability of the first moments).

On the other hand : the stability in the mean is identical with the stability of the undisturbed system

$$\frac{d\bar{x}}{dt} = A\bar{x} \,. \tag{2.47}$$

Thus, we have

Theorem 2.2 : The stability of the undisturbed system (2.47) is a necessary condition for the stability in mean square of the stochastically disturbed system (2.53).

2. 7. 4 Definitions of Stochastic Stability

Up to now we have investigated the stability of stochastic systems by taking averages and reducing to deterministic problems. Now we start with direct stochastic methods inevitable at least in the case of nonlinear stochastic systems.

According to the deterministic stability theory there is a great variety of stochastic stability definitions. The most important of them will be given here.

Consider the stochastic vector system

$$(2.57) \qquad dx = f(t,x)dt + \sigma(t,x)dw(t),$$

where $w(t)$ is a Wiener vector process with independent components $w_1(t), \ldots, w_r(t)$ and f and σ satisfy

$$(2.58) \qquad f(t,0) \equiv 0, \qquad \sigma(t,0) \equiv 0.$$

Due to (2.58) the origin $x = 0$ is an equilibrium position of the system (2.57).

Definition 1 : The equilibrium position $x = 0$ of the system (2.57) is said to be stable in probability, if the condition

$$(2.59) \qquad \lim_{x(t_0) \to 0} P\{\sup_{t > t_0} \|x(t)\| > \varepsilon\} = 0$$

holds for all $t > t_0 \geq 0$ and $\varepsilon > 0$.

The norm $\|x\|$ is defined as

$$\|x\| = (x_1^2 + \ldots + x_n^2)^{1/2}.$$

Remark : For many systems (e.g. for linear systems with constant coefficients) the stability in mean square implies the stability in probability.

Definition 2 : The origin $x = 0$ is asymptotically stable in probability if it is stable in probability and moreover :

$$\lim_{x(t_0)\to 0} P\{\lim_{t\to\infty} x(t) = 0\} = 1 . \qquad (2.60)$$

Definition 3 : The origin $x = 0$ is stable in the large if it is stable in probability and moreover

$$P\{\lim_{t\to\infty} x(t) = 0\} = 1 \qquad (2.61)$$

holds for all t_0 and $x(t_0)$.

2. 7. 5 Intuitive Approach to the Stochastic Liapunov Method

Given a deterministic system

$$\dot{x} = f(x) , \quad f(0) = 0$$

$$x^* = (x_1, \dots, x_n) , \quad f^* = (f_1, \dots, f_n)$$

the origin $x = 0$ is stable, if there exists a Liapunov function $v = v(x)$ with the properties

$$v(x) > 0 \quad \text{for} \quad x \neq 0 , \ v(0) = 0 \qquad (2.62)$$

$$\dot{v} = \sum_{i=1}^{n} \frac{\partial v}{\partial x_i} f_i = \frac{\partial v}{\partial x} f \leq 0 . \qquad (2.63)$$

This well known statement (direct method of Liapunov) becomes at once evident by geometric interpretation of conditions (2.62), (2.63). See[16].

In order to generalize this statement to the stochastic system

$$dx = f(x)dt + \sigma(x)dw$$

with a scalar Wiener process $w(t)$ consider a function $v(x)$ satisfying conditions (2.62). As in the deterministic case the decreasing of $v(x)$ along the trajectory $x(t)$ with time, i.e. $\dot{v} \leq 0$, implies a decreasing of x with time and thus the stability.

According to Itô formula (2.31) generalized to the case of a vector process x we have

$$dv(x) = \left(\frac{\partial v}{\partial x} f + \frac{1}{2}\sigma^* \frac{\partial^2 v}{\partial x^2}\sigma\right)dt + \frac{\partial v}{\partial x}\sigma dw$$

and thus

$$\dot{v} = \frac{\partial v}{\partial x} f + \frac{1}{2}\sigma^* \frac{\partial^2 v}{\partial x^2}\sigma + \frac{\partial v}{\partial x}\sigma \dot{w}\,.$$

Due to the white noise term on the right hand the condition $\dot{v} \leq 0$ can never be assured on some finite interval of time. To circumvent this difficulty it seems natural to replace the requirement $\dot{v} \leq 0$ by $E\dot{v} \leq 0$. Introducing the L-operator mentioned in 2.5.2. we have instead of (2.63) the condition

$$Lv \leq 0\,. \tag{2.64}$$

The strong formulation of this intuitively obtained stochastic analogon to the direct method of Liapunov runs as follows.

2.7.6 Strong Formulation of the Stochastic Liapunov Method

Given the stochastic system

$$dx = f(t,x)dt + \sigma(t,x)dw(t) \tag{2.65}$$

where $w(t)$ has independent components $w_1(t), \dots, w_r(t)$ and $f(t,0) \equiv 0$, $\sigma(t,0) \equiv 0$. Consider a function $v(t,x)$ with the properties

a) $v(t,0) = 0$

b) $v(t,0) > W(x) > 0$ for $x \neq 0$

c) $v(t,x)$ is continuous

d) v_t, v_x, v_{xx} exist and are continuous at least for $x \neq 0$.

Theorem 2.3 : Suppose, there exists a function $v(t,x)$ with properties a) to d) satisfying

$$Lv \leqslant 0 . \tag{2.66}$$

Then the equilibrium position $x = 0$ of eq. (2.65) is stable in probability.

The operator L is given by

$$L = \frac{\partial}{\partial t} + \sum_{i=1}^{n} f_i \frac{\partial}{\partial x_i} + \frac{1}{2}\sum_{i=1}^{n}\sum_{j=1}^{n}(\sigma\sigma^*)_{ij}\frac{\partial^2}{\partial x_i \partial x_j} . \tag{2.67}$$

2.7.7 The Operator L

The operator L can be obtained as follows. It is

$$dx = f\,dt + \sigma\,dw$$

$$v = v(t,x)$$

and

$$dv = v(t+dt, x_1+dx_1, \ldots, x_n+dx_n) - v(t,x_1,\ldots,x_n) =$$

$$= v_t dt + \sum_{i=1}^{n} v_{x_i} dx_i + \frac{1}{2}\sum_{i,j=1}^{n} v_{x_i x_j} dx_i dx_j + \ldots$$

$$= v_t dt + \sum_{i=1}^{n} v_{x_i}(f_i dt + \sum_{l=1}^{r} \sigma_{il} dw_l) +$$

$$+ \frac{1}{2}\sum_{i,j=1}^{n} v_{x_i x_j}(f_i dt + \sum_{l=1}^{r}\sigma_{il} dw_l)(f_j dt + \sum_{k=1}^{r}\sigma_{jk} dw_k) + \ldots$$

Neglecting in analogy to section 2.5 the terms with $(dt)^2$, $dt\,dw_m$, accounting the independence of w_l and w_k for $l \neq K$ and replacing $(dw_m)^2$ by dt we have

$$dv = (v_t + \sum_{i=1}^{n} v_{x_i} f_i + \frac{1}{2}\sum_{i,j=1}^{n} v_{x_i x_j}\sum_{k=1}^{r}\sigma_{ik}\sigma_{jk})dt +$$

$$+ \sum_{i=1}^{n}\sum_{l=1}^{r} v_{x_i}\sigma_{il} dw_l \,.$$

The operator L is defined by (see 2.5.2) :

$$Lv = v_t + \sum_{i=1}^{n} v_{x_i} f_i + \frac{1}{2}\sum_{i,j=1}^{n} v_{x_i x_j} \sum_{k=1}^{r} \sigma_{ik}\sigma_{jk} .$$

Now, the last sum in this expression is just the (i,j)-element of the matrix

$$(\sigma\sigma^*)_{ij} = \sum_{k=1}^{r} \sigma_{ik}\sigma_{jk} .$$

Thus, we have finally

$$L = \frac{\partial}{\partial t} + \sum_{i=1}^{n} f_i \frac{\partial}{\partial x_i} + \frac{1}{2}\sum_{i=1}^{n}\sum_{j=1}^{n} (\sigma\sigma^*)_{ij} \frac{\partial^2}{\partial x_i \partial x_j} . \quad (2.67)$$

2. 7. 8 An Important Special Case

Often the functions f and σ are independent of t

$$dx = f(x)dt + \sigma(x)dw(t) . \quad (2.68)$$

In this case the Liapunov function v is independent of t too and has to satisfy the conditions:

a') $v(0) = 0$

b') $v(x) > 0$ for $x \neq 0$

c') $v(x)$ is continuous

d') v_x, v_{xx} exist and are continuous at least for $x \neq 0$.

Theorem 2.4 : Suppose there exists a function $v(x)$ with properties a') to d') and

$$Lv \leq 0 \tag{2.69}$$

holds. Then the equilibrium position $x = 0$ of eq. (2.68) is stable in probability. The operator L is defined as

$$L = \sum_{i=1}^{n} f_i \frac{\partial}{\partial x_i} + \frac{1}{2} \sum_{i=1}^{n} \sum_{j=1}^{n} (\sigma\sigma^*)_{ij} \frac{\partial^2}{\partial x_i \partial x_j} . \tag{2.70}$$

If condition (2.69) can be sharpened to

$$Lv \leq W(x) \tag{2.71}$$

where $W(0) = 0$, $W(x) < 0$ for $x \neq 0$ the equilibrium position $x = 0$ is asymptotically stable in probability.

2.7.9 Attraction Domains

If the origin of a linear deterministic system is asymptotically stable every solution tends to the origin with $t \to \infty$. For non-linear systems this property holds not in general. Only solutions beginning in some bounded domain including an asymptotically stable equilibrium position tend to this position. The mentioned domains are called stability regions or domains of attraction and can be determined by means of modern Liapunov techniques.

These facts of the deterministic theory are extended to the non-linear stochastic differential equations in [3], [5].

2.7.10 Stability Check via Linearized Equation

The deterministic nonlinear differential equation

$$\dot{x} = f(x), \quad f(0) = 0, \tag{2.72}$$

and the corresponding linearized equation

$$\dot{x} = Ax \tag{2.73}$$

have (with exception of critical cases) the same stability behaviour in the neighborhood of the equilibrium position $x = 0$ [16]. This well known property makes it possible to reduce the stability investigation of deterministic nonlinear system (2.72) to the stability check of the linearized equation (2.73) (stability in first approximation).

The stochastic analogue of these statements is given by the following theorems [5] :

Theorem 2.5 : Consider the nonlinear system

$$(2.74)\qquad \begin{aligned} dx &= f(t,x)dt + \sum_{k=1}^{r} \sigma_k(t,x)dw_k(t) \ , \\ f(t,0) &= \sigma_k(t,0) = 0 \ , \end{aligned}$$

and the linear system with constant coefficients

$$(2.75)\qquad dx = Axdt + \sum_{k=1}^{r} B_k x dw_k \ .$$

Suppose, in the neighborhood of the equilibrium position $x = 0$ holds the estimate

$$|f(t,x) - Ax| + \sum_{k=1}^{r} |\sigma_k(t,x) - B_k x| < \gamma |x|$$

with a sufficiently small constant γ . Then : the asymptotic stability in probability of the equilibrium position $x = 0$ of eq. (2.75) implies the asymptotic stability in probability of the equilibrium position $x = 0$ of eq. (2.74).

Theorem 2.6 : Consider the nonlinear system

$$(2.76)\qquad dx = f(x)dt + \sum_{k=1}^{r} \sigma_k(x)dw_k$$

and the corresponding linearized system

$$(2.77)\qquad dx = Axdt + \sum_{k=1}^{r} B_k x dw_k$$

with bounded constant coefficients. The asymptotic stability in probability of eq. (2.77) implies the asymptotic stability in probability of eq. (2.76).

Remark 1 : The theory of stochastic differential equations and stochastic stability sketched in sections 2.3 to 2.7 can be extended to systems

$$dx = f(t,x)dt + \sigma(t,x)dq ,$$

where $q = q(t)$ is a Poisson process [3] . The most general stochastic differential equations are considered in [4] .

Remark 2 : The theorems 2.3 to 2.7 on stochastic stability remind of the deterministic theory. The extension of the deterministic instability theorems to stochastic systems is more complicated and is sometimes not possible at all [5] .

3. Orbiting Satellite Influenced by Gravity Gradient and Aerodynamic Torques

3.1 Librational Motion [11], [12], [13]

Let x, y, z denote the principal axes of inertia and A, B, C the corresponding principal moments of inertia of a rigid satellite in circular orbit (Fig. 3.1).

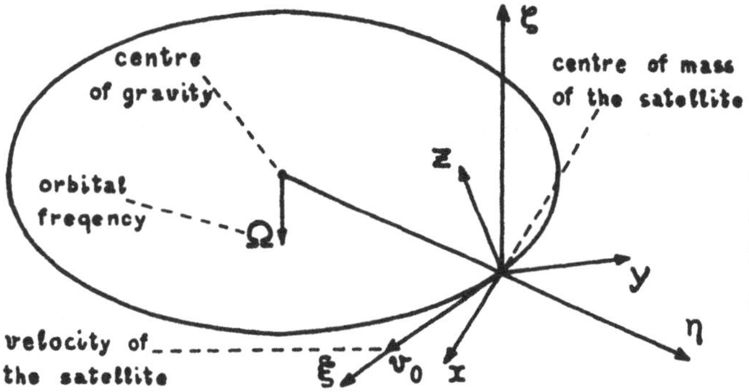

Fig. 3.1. Satellite-fixed x, y, z -axes and orbital axes ξ, η, ζ.

The orbital system ξ, η, ζ moving with the satellite is defined by the orbital tangent (ξ-axis), the satellite's radius vector relative to the centre of gravity (η-axis), and the normal of the orbital plane (ζ-axis). The motion of the satellite (xyz-tripod) relative to the orbital ξ, η, ζ-axes is described by the three Euler angles φ, ψ, ϑ (Fig. 3.2, see page 49). For small angles φ, ψ, ϑ the x, y, z-and the ξ, η, ζ-components of a vector can be transformed by equations

$$\begin{bmatrix} x \\ y \\ z \end{bmatrix} = T \begin{bmatrix} \xi \\ \eta \\ \zeta \end{bmatrix}, \qquad \begin{bmatrix} \xi \\ \eta \\ \zeta \end{bmatrix} = T^{-1} \begin{bmatrix} x \\ y \\ z \end{bmatrix} \tag{3.1}$$

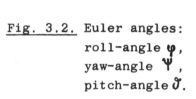

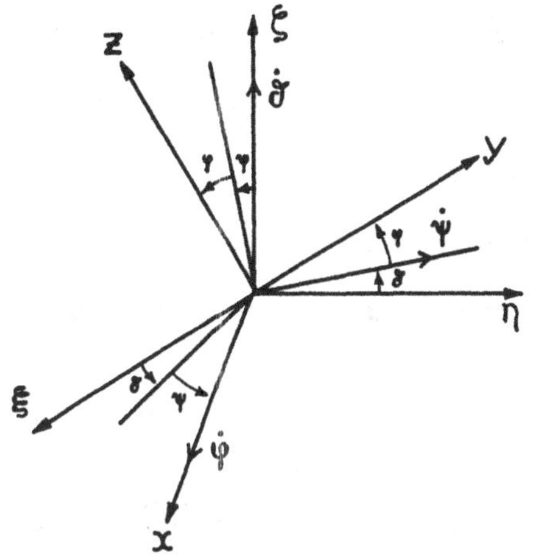

Fig. 3.2. Euler angles: roll-angle φ, yaw-angle Ψ, pitch-angle ϑ.

with the linearized transformation matrix

$$T = \begin{bmatrix} 1 & \vartheta & -\Psi \\ -\vartheta & 1 & \varphi \\ \Psi & -\varphi & 1 \end{bmatrix}. \tag{3.2}$$

The librational motion of the satellite about its centre of mass is described by the Euler equation

$$\frac{d\bar{D}}{dt} + \bar{\omega} \times \bar{D} = \bar{M} \tag{3.3}$$

where $\bar{D}$ is the moment of momentum of the satellite, $\bar{\omega}$ the absolute angular velocity of the satellite and $\bar{M}$ the external torque.

For small deviations φ, Ψ, ϑ the x, y, z - components of $\bar{\omega}$ are :

$$\omega_x = \dot{\varphi} + \Omega\Psi$$
(3.4)
$$\omega_y = \dot{\Psi} - \Omega\varphi$$
$$\omega_z = \dot{\vartheta} - \Omega$$

where Ω is the constant orbital frequency (Fig. 3.2).

Inserting the relations (3.4) in eq. (3.3) and neglecting terms of higher order we get the linearized equations of the librational motion for a satellite in circular orbit :

$$A\ddot{\varphi} + \Omega^2(C - B)\varphi + \Omega(A + B - C)\dot{\Psi} = M_x$$
(3.5)
$$B\ddot{\Psi} + \Omega^2(C - A)\Psi - \Omega(A + B - C)\dot{\varphi} = M_y$$
$$C\ddot{\vartheta} = M_z \,.$$

Here, M_x, M_y, M_z are the components of the external torques about the x, y, z-axes.

3.2 Gravity Gradient Torques [11], [12], [13]

The gravity gradient torques acting on a satellite are

$$M_{gx} = 3\Omega^2(C - B)\gamma'\gamma''$$
(3.6)
$$M_{gy} = 3\Omega^2(A - C)\gamma\gamma''$$
$$M_{gz} = 3\Omega^2(B - A)\gamma\gamma' ,$$

where $\gamma, \gamma', \gamma''$ denote the direction cosines of the η-axis in the x, y, z-system.

For small angles φ, Ψ, ϑ the formulas (3.6) can be replaced by

$$\begin{aligned} M_{gx} &= -3\Omega^2(C - B)\varphi \\ M_{gy} &= 0 \\ M_{gz} &= -3\Omega^2(A - B)\vartheta \, . \end{aligned} \tag{3.7}$$

3.3 Aerodynamic Torques [12], [14], [15], [20]

In general case the vector of the aerodynamic torque is given by

$$\bar{M}_a = \frac{1}{2}\varrho v^2[\bar{f}(\varphi, \Psi, \vartheta) + K\bar{\omega}] \, . \tag{3.8}$$

Here, ϱ is the local atmospheric density and v is the satellite velocity relative to the atmosphere. Neglecting the motion of the atmosphere with respect to the earth, v is identical to the satellite's velocity v_0 and thus $v = v_0$ = const for circular orbits. The vector-function $\bar{f}$ and the coefficient matrix K depend on satellite geometry. The first term in the brackets of eq. (3.8) has a restoring and the second term a dissipative nature.

The density ϱ depends on day time, on

the earth-moon configuration, on latitude, on the activity of the sun, on earth's magnetic field etc. The fluctuations of the density increase with altitude and cause density changes of about three orders of magnitude. Thus, the density must be considered as a stochastic process. An appropriate model for ϱ is

$$\varrho = \varrho_0(1 + \delta\dot{w}) \tag{3.9}$$

where ϱ_0 denotes some mean value depending on altitude only (ϱ_0 = const in circular orbit), δ is a constant and $\dot{w}$ is a white noise.

3.4 Stochastic Differential Equation of the Librational Motion [12] [15]

The librational motion of a satellite influenced by gravity gradient and aerodynamic torques is described by eq. (3.3) with $\bar{M} = \bar{M}_g + \bar{M}_a$:

$$\frac{d\bar{D}}{dt} + \bar{\omega}\times\bar{D} = \bar{M}_g + \frac{1}{2}\varrho v^2[\bar{f}(\varphi,\Psi,\vartheta) + K\bar{\omega}] \tag{3.10}$$

where $\bar{M}_g$ is given by eq. (3.6). Introducing for the density ϱ the model (3.9), a stochastic differential equation results :

$$\frac{d\bar{D}}{dt} = \left[-\bar{\omega}\times\bar{D} + \bar{M}_g + \frac{1}{2}\varrho_0 v^2(\bar{f} + K\bar{\omega})\right] + \frac{1}{2}\varrho_0 v^2(\bar{f} + K\bar{\omega})\delta\dot{w} \; . \tag{3.11}$$

In the general case of a three-dimensional motion $\varphi = \varphi(t)$, $\Psi = \Psi(t)$, $\vartheta = \vartheta(t)$ the investigation of the highly nonlinear vector equation (3.11) becomes very cumbersome and is beyond the scope of this introductory textbook. Therefore, in this paper only the plane pitch motion of a symmetrical satellite in circular orbit (Fig. 3.3) will be discussed in detail.

Fig. 3.3. To the plane pitch motion of a symmetrical satellite in a circular orbit.

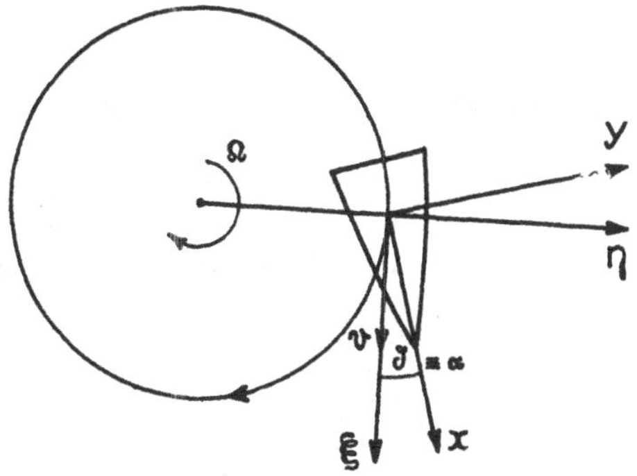

Suppose : 1. the satellite is in circular orbit, i.e. $v = v_0 = \text{const}$, $\varrho_0 = \text{const}$; 2. roll-and yaw-angles are at equilibrium, i.e. $\varphi = 0$, $\Psi = 0$; 3. the x-axis is an axis of symmetry of the satellite and it is $A < B = C$. In this case it can easily be verified (Fig. 3.2 and fig. 3.1) that the following relations hold

$$\bar{\omega} = \begin{bmatrix} 0 \\ 0 \\ \omega_z \end{bmatrix}, \quad \bar{D} = \begin{bmatrix} 0 \\ 0 \\ C\omega_z \end{bmatrix}, \quad \begin{bmatrix} \gamma \\ \gamma' \\ \gamma'' \end{bmatrix} = \begin{bmatrix} \sin\vartheta \\ \cos\vartheta \\ 0 \end{bmatrix}. \quad (3.12)$$

The absolute angular velocity ω_z is (Fig. 3.1) :

$$\omega_z = \dot{\vartheta} - \Omega . \tag{3.13}$$

The gravity gradient torque (3.6) reduces in the considered case to

$$\bar{M}_g = \begin{bmatrix} 0 \\ 0 \\ \frac{3}{2}\Omega^2(B - A)\sin 2\vartheta \end{bmatrix} . \tag{3.14}$$

Furthermore, due to the supposed symmetry of the satellite, the vectors $\bar{f}(\varphi, \Psi, \vartheta)$ and $K\bar{\omega}$ can be described with a great degree of accuracy by

$$\bar{f} = \begin{bmatrix} 0 \\ 0 \\ -F_0 d c_1 g(\vartheta) \end{bmatrix} , \quad K\bar{\omega} = \begin{bmatrix} 0 \\ 0 \\ -F_0 d c_2 \dot{\vartheta} \end{bmatrix} , \tag{3.15}$$

where F_0 and d denote the reference area and the reference length of the satellite, respectively ; c_1 and c_2 are the static and the dynamic aerodynamic moment coefficients, respectively. For the considered symmetric satellite the function $g(\vartheta)$ has the obvious properties :

$$g(0) = 0 , \quad g(\vartheta + 2\pi) = g(\vartheta) , \quad g(-\vartheta) = - g(\vartheta) . \tag{3.16}$$

Moreover, by appropriate normalization of the coefficient c_1 the inequality

$$|g(\vartheta)| \leq |\vartheta| \tag{3.17}$$

can always be supposed. Usually the approximations $g(\vartheta) = \sin\vartheta$ and, for small angles, $g(\vartheta) = \vartheta$ are used. Sometimes the second harmonic of $g(\vartheta)$ is also taken into account. Two typical curves $g(\vartheta)$ are shown in Fig. 3.4.

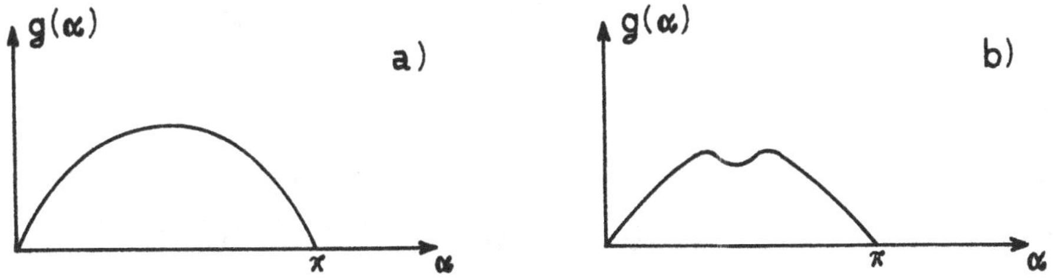

Fig. 3.4. To dependence of the aerodynamic torque on satellite's geometry: a) sphere, b) cone.

Incerting the expressions (3.12) to (3.15) into eq. (3.11), regarding the symmetry (B = C) of the satellite and using the abbreviations

$$Q = \varrho_0 v_0^2 F_0 d c_1 / 2C \,, \qquad (3.18)$$

$$\eta = c_2/c_1 \qquad (3.19)$$

$$Q\ell = \frac{3}{2}\Omega^2\left(1 - \frac{A}{C}\right) \qquad (3.20)$$

the pitch equation becomes

$$\ddot{\vartheta} + Q\left[g(\vartheta) + \eta\dot{\vartheta}\right] - Q\ell\sin 2\vartheta + Q\delta\left[g(\vartheta) + \eta\dot{\vartheta}\right]\dot{w} = 0 \,. \qquad (3.21)$$

This equation can be simplified if the normalized time

$$\tau = \sqrt{Q}\,t \qquad (3.22)$$

and the new constants

$$(3.23) \qquad \delta_0^2 = \sqrt{Q}\delta^2, \quad \eta_0 = \sqrt{Q}\eta$$

are introduced. Then eq. (3.21) can be written as

$$(3.24) \quad \vartheta'' + g(\vartheta) + \eta_0\vartheta' - \ell\sin 2\vartheta + \delta_0[g(\vartheta) + \eta_0\vartheta']w' = 0$$

where $' \equiv d/d\tau$.

Thus, the pitch motion of a symmetric satellite influenced by gravity gradient and aerodynamic torques is described by eq. (3.24).

For small angles ϑ eq. (3.24) can be linearized replacing $\sin 2\vartheta$ by 2ϑ and $g(\vartheta)$ by $\mu\vartheta$, respectively. Then small pitch oscillations are described by

$$(3.25) \quad \vartheta'' + \eta_0(1 + \delta_0 w')\vartheta' + (\mu - 2\ell + \delta_0\mu w')\vartheta = 0$$

where

$$(3.26) \qquad \mu = \left.\frac{dg(\vartheta)}{d\vartheta}\right|_{\vartheta = 0}$$

and $\mu \leq 1$ in consequence of eq. (3.17).

Introducing the state variables $x_1 = \vartheta$ and $x_2 = \vartheta'$ the eqs. (3.24) and (3.25) can be written as stochastic Itô systems. To the nonlinear case (3.24) corresponds the system

$$d\begin{bmatrix} x_1 \\ x_2 \end{bmatrix} = \begin{bmatrix} x_2 \\ -g(x_1) + l\sin 2x_1 - \eta_0 x_2 \end{bmatrix} dt + \begin{bmatrix} 0 \\ -\delta_0(g(x_1) + \eta_0 x_2) \end{bmatrix} dw \,. \tag{3.27}$$

In the linear case (3.25) the corresponding Itô system is

$$dx = Fx d\tau + Gx dw(\tau) \tag{3.28}$$

with

$$x = \begin{bmatrix} x_1 \\ x_2 \end{bmatrix}, \; F = \begin{bmatrix} 0 & 1 \\ -(\mu - 2l) & -\eta_0 \end{bmatrix}, \; G = \begin{bmatrix} 0 & 0 \\ -\delta_0\mu & -\delta_0\eta_0 \end{bmatrix}. \tag{3.29}$$

Remark 1 : Due to the assumption $A < C$, the gravity gradient torque is destabilizing.

Remark 2 : In the considered case of the plane pitch motion of a symmetric satellite in circular orbit, the pitch angle ϑ is identical with the angle of attack α . Neglecting the motion of the atmosphere with respect to the earth, α is defined as the angle between the satellite's axis of symmetry and the satellite's velocity (Fig. 3.3).

3.5 Stability of the Pitch Motion

In this section the stability of the pitch motion will be investigated first by means of the covariance equation and then by Liapunov function techniques. Finally, the results are discussed.

3.5.1 Stability Conditions via Covariance Equation

The covariance matrix corresponding to the solution of eq. (3.28) is

$$E(xx^*) = P = \begin{bmatrix} p_{11} & p_{12} \\ p_{21} & p_{22} \end{bmatrix}, \quad p_{21} = p_{12}.$$

The matrix P satisfies the equation

$$P' = FP + PF^* + GPG^*, \tag{2.51}$$

where the matrices F and G are given by the expressions (3.29).

It is

$$FP = \begin{bmatrix} 0 & 1 \\ -(\mu - 2\ell) & -\eta_0 \end{bmatrix} \begin{bmatrix} p_{11} & p_{12} \\ p_{12} & p_{22} \end{bmatrix} =$$

$$= \begin{bmatrix} p_{12} & p_{22} \\ -(\mu - 2\ell)p_{11} - \eta_0 p_{12} & -(\mu - 2\ell)p_{12} - \eta_0 p_{22} \end{bmatrix}$$

and

$$FP + PF^* = FP + (FP)^* =$$

$$= \begin{bmatrix} 2p_{12} & -(\mu - 2\ell)p_{11} - \eta_0 p_{12} + p_{22} \\ -(\mu - 2\ell)p_{11} - \eta_0 p_{12} + p_{22} & -2(\mu - 2\ell)p_{12} - 2\eta_0 p_{22} \end{bmatrix} .$$

The matrix GPG^* is

$$GPG^* = \begin{bmatrix} 0 & 0 \\ -\delta_0\mu & -\delta_0\eta_0 \end{bmatrix} \begin{bmatrix} p_{11} & p_{12} \\ p_{12} & p_{22} \end{bmatrix} \begin{bmatrix} 0 & -\delta_0\mu \\ 0 & -\delta_0\mu_0 \end{bmatrix} =$$

$$= \begin{bmatrix} 0 & 0 \\ -\delta_0\mu p_{11} - \delta_0\eta_0 p_{12} & -\delta_0\mu p_{12} - \delta_0\eta_0 p_{22} \end{bmatrix} \begin{bmatrix} 0 & -\delta_0\mu \\ 0 & -\delta_0\eta_0 \end{bmatrix} =$$

$$= \begin{bmatrix} 0 & 0 \\ 0 & \delta_0^2\mu^2 p_{11} + 2\delta_0^2\mu\eta_0 p_{12} + \delta_0^2\eta_0^2 p_{22} \end{bmatrix} .$$

Thus, the right side of the covariance equation (2.51) is

$$FP + PF^* + GPG^* =$$

$$= \begin{bmatrix} 2p_{12} & -(\mu - 2\ell)p_{11} - \eta_0 p_{12} + p_{22} \\ -(\mu - 2\ell)p_{11} - \eta_0 p_{12} + p_{22} & \delta_0^2\mu^2 p_{11} + [2\delta_0^2\mu\eta_0 - 2(\mu - 2\ell)]p_{12} + [\delta_0^2\eta_0^2 - 2\eta_0]p_{22} \end{bmatrix} .$$

Now, the matrix covariance equation (2.51) can be written explicitly as the linear system

$$\frac{d}{d\tau}\begin{bmatrix} p_{11} \\ p_{12} \\ p_{22} \end{bmatrix} = N \begin{bmatrix} p_{11} \\ p_{12} \\ p_{22} \end{bmatrix} \tag{3.30}$$

with the system matrix

$$N = \begin{bmatrix} 0 & 2 & 0 \\ -(\mu - 2\ell) & -\eta_0 & 1 \\ \delta_0^2\mu^2 & 2\delta^2\mu\eta_0 - 2(\mu - 2\ell) & \delta_0^2\eta_0^2 - 2\eta_0 \end{bmatrix} . \tag{3.31}$$

The characteristic equation corresponding to the system (3.30) or to the matrix (3.31) is

$$s^3 + a_1 s^2 + a_2 s + a_3 = 0 \tag{3.32}$$

with

$$\begin{aligned} a_1 &= \eta_0(3 - \delta_0^2\eta_0) \\ a_2 &= 4(\mu - 2\ell) - 2\delta_0^2\mu\eta_0 + 2\eta_0^2 - \delta_0^2\eta_0^3 \\ a_3 &= 2[(\mu - 2\ell)(2\eta_0 - \delta_0^2\eta_0^2) - \delta_0^2\mu^2] . \end{aligned} \tag{3.33}$$

The well known Hurwitz stability conditions for eq. (3.32) are [16]

$$a_1 > 0 , \quad a_2 > 0 , \quad a_3 > 0 \tag{3.34}$$

$$a_1 a_2 - a_3 > 0 . \tag{3.35}$$

Suppose first, the inequality $\mu - 2\ell < 0$ holds. In this case the coefficient a_3 can be positive only if

$2\eta_0 - \delta_0^2\eta_0^2 < 0$ or $\delta_0^2 > 2/\eta_0$. Assume it is

$$\delta_0^2 = \frac{2+\varepsilon}{\eta_0}, \quad \varepsilon > 0 .$$

Then

$$\begin{aligned} a_2 &= 4(\mu - 2\ell) - 2\mu\eta_0\frac{2+\varepsilon}{\eta_0} + 2\eta_0^2 - \eta_0^3\frac{2+\varepsilon}{\eta_0} = \\ &= 4\mu - 8\ell - 4\mu - 2\mu\varepsilon + 2\eta_0^2 - 2\eta_0^2 - \varepsilon\eta_0^2 = \\ &= -8\ell - 2\mu\varepsilon - \varepsilon\eta_0^2 < 0 , \end{aligned}$$

which contradicts condition (3.34). Thus, the condition

$$\mu - 2\ell > 0 \tag{3.36}$$

is necessary for the stability of the moments. Physically this means, that the pitch motion disturbed by noise ($\delta_0 \neq 0$) can have stable moments only if the undisturbed motion ($\delta_0 \equiv 0$) is stable (see theorem 2.2) .

Regarding the expressions (3.33) and the necessary condition (3.36), the stability conditions (3.34) can be written as

$$\delta_0^2 < \frac{3}{\eta_0}$$

$$\delta_0^2 < \frac{2\eta_0^2 + 4(\mu - 2\ell)}{\eta_0^3 + 2\mu\eta_0}$$

$$\delta_0^2 < \frac{2\eta_0(\mu - 2l)}{\eta_0^2(\mu - 2l) + \mu^2}$$

or as

$$(3.37) \quad \delta_0^2 < \min\left\{\frac{3}{\eta_0}, \frac{2\eta_0^2 + 4(\mu - 2l)}{\eta_0^3 + 2\mu\eta_0}, \frac{2\eta_0(\mu - 2l)}{\eta_0^2(\mu - 2l) + \mu^2}\right\} .$$

In consequence of

$$\frac{2\eta_0(\mu - 2l)}{\eta_0^2(\mu - 2l) + \mu^2} = \frac{2}{\eta_0 + \dfrac{\mu^2}{\eta_0(\mu - 2l)}} < \frac{3}{\eta_0}$$

the condition (3.37) reduces to

$$\delta_0^2 < \min\left\{\frac{2\eta_0^2 + 4(\mu - 2l)}{\eta_0^3 + 2\mu\eta_0}, \frac{2\eta_0(\mu - 2l)}{\eta_0^2(\mu - 2l) + \mu^2}\right\} .$$

Comparing directly both terms in the brackets it follows:

$$(3.38) \qquad \delta_0^2 < \delta_{0\,max}^2 = \frac{2\eta_0(\mu - 2l)}{\eta_0^2(\mu - 2l) + \mu^2} .$$

The stability condition (3.35) remains to be checked. Inserting eqs. (3.33) into condition (3.35) we get :

$$(3.39) \quad \begin{aligned} a_1 a_2 - a_3 = h(\delta_0^2) = {} & \eta_0^3(\eta_0^2 + 2\mu)\delta_0^4 - [5\eta_0^4 + 4(\mu - 2l)\eta_0^2 - \\ & - 2\mu^2]\delta_0^2 + 2\eta_0[3\eta_0^2 + 4(\mu - 2l)] . \end{aligned}$$

For $5\eta_0^4 + 4(\mu - 2l)\eta_0^2 - 2\mu^2 < 0$ the stability condition (3.35) is satisfied. Suppose,

$$5\eta_0^4 + 4(\mu - 2l)\eta_0^2 - 2\mu^2 > 0 .$$

The function $h = h(\delta_0^2)$ is quadratic in δ_0^2 describes a parabola and has the properties $1.\ h(0) > 0$, $2.\ h(\delta_{0max}^2) > 0$ (a_3 vanishes for $\delta_0^2 = \delta_{0max}^2$ and a_1, a_2 are both positive for $\delta_0^2 = \delta_{0max}^2$, $3.\ h(\infty) = +\infty$, $4.\ h = h(\delta_0^2)$ has its minimum for

$$\delta_{01}^2 = \frac{5\eta_0^4 + 4(\mu - 2l)\eta_0^2 - 2\mu^2}{2\eta_0^3(\eta_0^2 + 2\mu)} . \tag{3.40}$$

Thus, four positions of the parabola $h = h(\delta_0^2)$ are possible (Fig. 3.5).

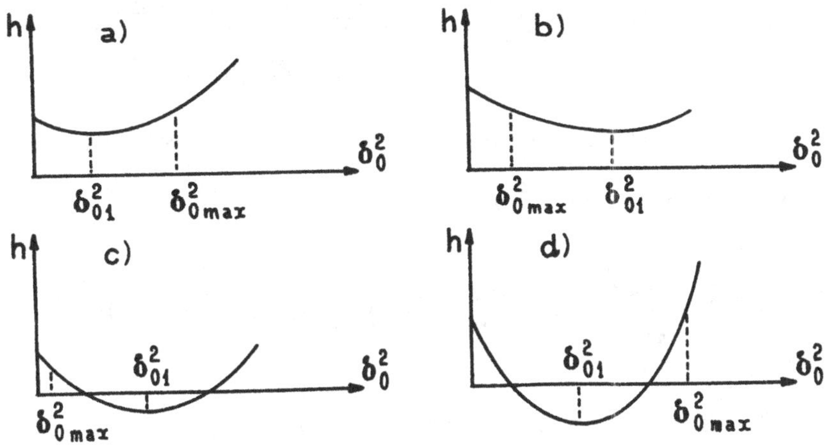

Fig. 3.5. To the stability condition (3.35). In the text, case d) is shown to be impossible.

If the parabola has the positions a), b) or c), the condition (3.35) is fulfilled for every δ_0^2 satisfying condition (3.38). Let us show that the fourth position d) is not possible. For this purpose it is sufficient to show that for arbitrary positive constants η_0, ℓ, $\mu > 2\ell$ at least one of the following two inequalities holds :

(3.41) $$h(\delta_{01}^2) > 0$$

(3.42) $$\delta_{0\,max}^2 < \delta_{01}^2 .$$

Taking into account eq. (3.40), the conditions (3.41), (3.42) can be replaced by

$$\left[5\eta_0^4 + 4(\mu - 2\ell)\eta_0^2 - 2\mu^2\right]^2 < 8\eta_0^4(\eta_0^2 + 2\mu)\left[3\eta_0^2 + 4(\mu - 2\ell)\right]$$

$$\left[5\eta_0^4 + 4(\mu - 2\ell)\eta_0^2 - 2\mu^2\right]^2 > \frac{16\eta_0^8(\mu - 2\ell)^2(\eta_0^2 + 2\mu)^2}{\left[(\mu - 2\ell)\eta_0^2 + \mu^2\right]^2} .$$

At least one of these inequalitites is satisfied for arbitrary positive η_0, ℓ, $\mu > 2\ell$ if

$$\frac{16\eta_0^8(\mu - 2\ell)^2(\eta_0^2 + 2\mu)^2}{\left[(\mu - 2\ell)\eta_0^2 + \mu^2\right]^2} \leq 8\eta_0^4(\eta_0^2 + 2\mu)\left[3\eta_0^2 + 4(\mu - 2\ell)\right] .$$

By simple algebra this last inequality can be reduced to

$$(\mu - 2\ell)\eta_0^6 + 2(\mu - 2\ell)\left[2\mu^2 + (\mu - 2\ell)^2 + 4\ell^2\right]\eta_0^4 +$$
$$+ \mu^2\left[8(\mu - 2\ell)^2 + 3\mu^2\right]\eta^2 + 4\mu^4(\mu - 2\ell) \geq 0 ,$$

q.e.d.

Thus, we have stated : The second (and in consequence the first (see theorem 2.1) moments of the stochastically disturbed linearized pitch motion (3.28), (3.29) are stable if and only if the intensity of the white noise satisfies

$$\delta_0^2 < \delta_{0max}^2 = \frac{2\eta_0(\mu - 2\ell)}{\eta_0^2(\mu - 2\ell) + \mu^2} . \qquad (3.38)$$

Introducing as a new parameter

$$q^2 = \frac{\mu^2}{\mu - 2\ell}$$

condition (3.38) reads as

$$\delta_0^2 < \frac{2\eta_0}{\eta_0^2 + q^2} .$$

In fig. 3.6 the stability domain is shown for different values of q^2. The stability domain decreases with increasing q^2. If the satellite has a spherical mass distribution (A = B = C) the gravity gradient torque is zero and $\ell = 0$. In this case $q^2 = \mu \leq 1$. For $g(x_1) = \sin x_1$ we have $\mu = 1$. The corresponding stability domain is shown in Fig. 3.7.

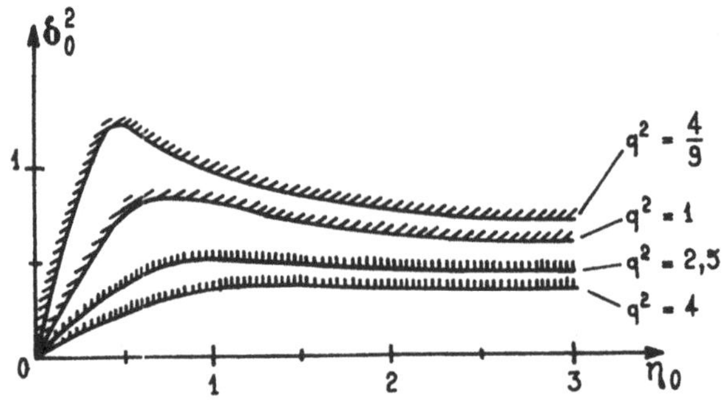

Fig. 3.6. The stability domains of the moments for the linearized pitch motion.

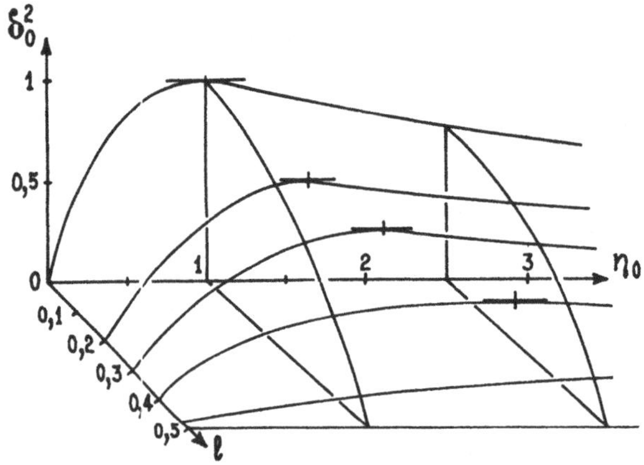

Fig. 3.7. Stability domain of the moments for the case $\mu = 1$.

Remark : As the stochastic system (3.28), (3 .29) is equivalent to the one second order equation (3.25) the condition (3.38) can be obtained directly from eq. (3.25) using the criteria given in [5] and mentioned in section 2. Here, the longer way via covariance equation was prefered as the more typical one, because the reduction of a system to one equation of higher order is often undesirable [17] and sometimes not possible at all.

3.5.2 Stability Conditions via Liapunov Techniques [15]

For simplicity reasons the stochastic stability of the non-linear pitch motion (3.27) will be considered only for the case $g(x_1) = \sin x_1$. Then, the equation of motion is

$$d\begin{bmatrix} x_1 \\ x_2 \end{bmatrix} = \begin{bmatrix} x_2 \\ -\sin x_1 + l\sin 2x_1 - \eta_0 x_2 \end{bmatrix} d\tau + \begin{bmatrix} 0 \\ -\delta_0(\sin x_1 + \eta_0 x_2) \end{bmatrix} dw \; . \qquad (3.43)$$

For the investigation of the stochastic stability of system (3.43) a Liapunov function has to be constructed. It seems reasonable to use the deterministic experience and to suppose this function to be a linear combination of a quadratic form in state variables and of the nonlinearities, i.e. :

$$\begin{aligned} v &= ax_1^2 + bx_1x_2 + x_2^2 + d\int_0^{x_1} \sin y\, dy + e\int_0^{x_1} \sin 2y\, dy = \\ &= ax_1^2 + bx_1x_2 + x_2^2 + d(1 - \cos x_1) + \frac{1}{2}e(1 - \cos 2x_1) \; . \end{aligned} \qquad (3.44)$$

Using well known trigonometric relations the Liapunov function becomes

$$v(x) = ax_1^2 + bx_1x_2 + x_2^2 + 2d\sin^2\frac{x_1}{2} + e\sin^2 x_1 \; . \qquad (3.45)$$

According to the definition (2.70) we need the following relations for the calculation of Lv :

$$\frac{\partial v}{\partial x_1} = 2ax_1 + bx_2 + 2d\sin\frac{x_1}{2}\cos\frac{x_1}{2} + 2e\sin x_1\cos x_1$$

$$\frac{\partial v}{\partial x_2} = bx_1 + 2x_2$$

$$\begin{bmatrix} 0 \\ -\delta_0(\sin x_1 + \eta_0 x_2) \end{bmatrix} \left[0,\ -\delta_0(\sin x_1 + \eta_0 x_2)\right] = \begin{bmatrix} 0 & 0 \\ 0 & \delta_0^2(\sin x_1 + \eta_0 x_2)^2 \end{bmatrix}$$

$$\frac{\partial^2 v}{\partial x_2^2} = 2\ .$$

Using eq. (2.70) and the relations above we get :

$$\begin{aligned} Lv = {} & 2ax_1x_2 + bx_2^2 + 2dx_2\sin\frac{x_1}{2}\cos\frac{x_1}{2} + 2ex_2\sin x_1\cos x_1 - \\ & - bx_1\sin x_1 + b\ell x_1\sin 2x_1 - b\eta_0 x_1x_2 - 2x_2\sin x_1 + \\ & + 2\ell x_2\sin 2x_1 - 2\eta_0 x_2^2 + \delta_0^2\sin^2 x_1 + 2\delta_0^2\eta_0 x_2\sin x_1 + \\ & + \delta_0^2\eta_0^2 x_2^2 = \\ = {} & (2a - b\eta_0)x_1x_2 - (2\eta_0 - b - \delta_0^2\eta_0^2)x_2^2 + \\ & + \left[d - 2 + 2\delta_0^2\eta_0 + (4\ell + 2e)\cos x_1\right]x_2\sin x_1 - \\ & - (b - 2b\ell\cos x_1 - \delta_0^2\frac{\sin x_1}{x_1})x_1\sin x_1\ . \end{aligned}$$

To assure the definiteness of this expression the free constants a, d, e have to be fixed as follows

$$2a - b\eta_0 = 0\ ,$$

$$4\ell + 2e = 0 ,$$

$$d - 2 + 2\delta_0^2\eta_0 = 0 .$$

Now the function $v(x)$ given by eq. (3.45) reads as

$$v(x) = \frac{1}{2}b\eta_0 x_1^2 + bx_1x_2 + x_2^2 + + 4\sin^2\frac{x_1}{2}(1 - \eta_0\delta_0^2 - 2\ell\cos^2\frac{x_1}{2}) \tag{3.46}$$

and the function $Lv(x)$ becomes

$$Lv(x) = -(2\eta_0 - b - \delta_0^2\eta_0^2)x_2^2 - - (b - 2b\ell\cos x_1 - \delta_0^2\frac{\sin x_1}{x_1})x_1\sin x_1 . \tag{3.47}$$

For positive definiteness of v the conditions

$$0 < b < 2\eta_0 \tag{3.48}$$

and

$$\delta_0^2 < \frac{1 - 2\ell}{\eta_0} \tag{3.49}$$

are sufficient.

The function $Lv(x)$ can be maximized by $-W(x)$

$$Lv(x) \leq -W(x)$$

with

$$W(x) = (2\eta_0 - b - \delta_0^2\eta_0^2)x_2^2 + (b - 2b\ell - \delta_0^2)x_1 \sin x_1 .$$

The function $W(x)$ is positive definite if the intensity of the white noise satisfies the conditions :

$$\delta_0^2 < \frac{2\eta_0 - b}{\eta_0^2} \tag{3.50}$$

$$\delta_0^2 < b(1 - 2\ell) . \tag{3.51}$$

Due to the theorem 2.4, the asymptotic stability in probability of the equilibrium position $x = 0$ is guaranteed, if conditions (3.48) to (3.51) are fulfilled.

As the stability conditions still depend on the unknown constant b the question arises how to fix this constant. The simplest way is to choose b in the middle of the interval (3.48), i.e. to fix $b = \eta_0$. In this case the stability conditions (3.49) to (3.51) reduce to

$$\delta_0^2 < \min\left\{\eta_0(1 - 2\ell) , \frac{1 - 2\ell}{\eta_0}\right\} . \tag{3.52}$$

This condition was first given in [15]. A better choice of b can be based on the inspection of conditions (3.50) and (3.51). The stability domains corresponding to the conditions (3.50) and (3.51) in the plane (b , δ_0^2) are shown in fig. 3.8.

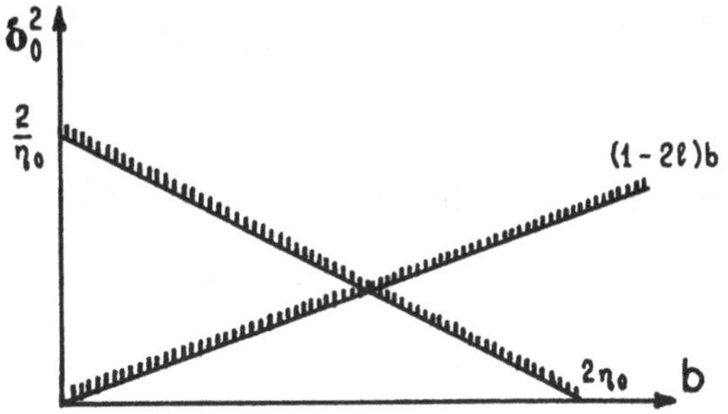

Fig. 3.8.

The conditions (3.50) and (3.51) in dependence on b.

The widest bound for δ_0^2 is obtained at the intersection of the straight lines

$$\delta_0^2 = -\frac{1}{\eta_0^2} b + \frac{2}{\eta_0}$$

and

$$\delta_0^2 = (1 - 2\ell) b$$

i.e. for

$$b = \frac{2\eta_0}{1 + \eta_0^2(1 - 2\ell)} < 2\eta_0 . \qquad (3.53)$$

Inserting the value (3.53) the expression for $W(x)$ becomes

$$W(x) = (2\eta_0 - \frac{2\eta_0}{1 + \eta_0^2(1 - 2\ell)} - \delta_0^2\eta_0^2)x_2^2 +$$

$$+ \left[(1 - 2\ell)\frac{2\eta_0}{1 + \eta_0^2(1 - 2\ell)} - \delta_0^2\right]x_1 \sin x_1 =$$

$$= \left[\frac{2\eta_0(1 - 2\ell)}{1 + \eta_0^2(1 - 2\ell)} - \delta_0^2\right](\eta_0^2 x_2^2 + x_1 \sin x_1) .$$

It is $W(0) = 0$ and $W(x) > 0$ for

$$(3.54) \qquad \delta_0^2 < \frac{2\eta_0(1 - 2\ell)}{1 + \eta_0^2(1 - 2\ell)}$$

Inserting the value (3.53) of b into eq. (3.46) and rearranging the terms we obtain :

$$v(x) = \left(\frac{\eta_0 x_1}{1 + (1 - 2\ell)\eta_0^2} + x_2\right)^2 + x_1^2\left[\frac{(1 - 2\ell)\eta_0^4}{(1 + (1 - 2\ell)\eta_0^2)^2} + \right.$$
$$\left. + \left(\frac{\sin\frac{x_1}{2}}{\frac{x_1}{2}}\right)^2 (1 - \eta_0\delta_0^2 - 2\ell\cos^2\frac{x_1}{2})\right] .$$

Taking into account the inequalities (3.54) the function $v(x)$ can be minimized as follows

$$v(x) \geq \left(\frac{\eta_0 x_1}{1 + (1 - 2\ell)\eta_0^2} + x_2\right)^2 + \left[\frac{(1 - 2\ell)\eta_0^4}{(1 + (1 - 2\ell)\eta_0^2)^2} + \right.$$
$$\left. + \frac{1 - (1 - 2\ell)\eta_0^2}{1 + (1 - 2\ell)\eta^2} - 2\ell\right]x_1^2$$

or

$$v(x) \geq \left(\frac{\eta_0 x_1}{1 + (1 - 2\ell)\eta_0^2} + x_2\right)^2 + \frac{(1 - 2\ell)(2\ell\eta_0^2 - 1)^2}{(1 + (1 - 2\ell)\eta_0^2)^2} x_1^2 .$$

i.e. the function $v(x)$ is positive definite for δ_0 satisfying condition (3.54). Thus, we have shown :
The equilibrium position $x_1 = x_2 = 0$ of the nonlinear

pitch motion (3.27) with $g(x_1) = \sin x_1$ (or (3.43))is asymptotically stable in probability if the intensity of the white noise satisfies condition (3.54).

Fig. 3.9 gives the stability domains corresponding to conditions (3.52) and (3.54) for $\ell = 0.3$.

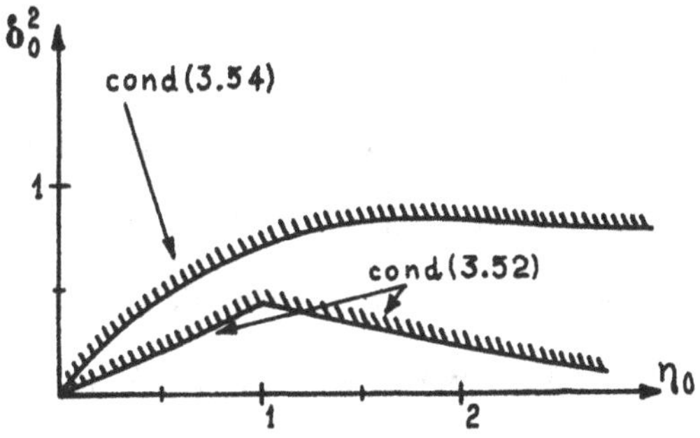

Fig. 3.9. The stability domains corresponding to the conditions (3.52) and (3.54), respectively, for $\ell = 0{,}3$.

Obviously, by a thorough discussion of the definiteness conditions and by a proper choice of b the sufficient stability condition (3.52) given in [15] can be considerably weakened.

Remark 1 : The system (3.43) has the equilibrium positions

$$x_1 = 0 \pm 2K\pi , \quad x_2 = 0 , \quad K = 1,2,\ldots \qquad (3.55)$$

$$x_1 = \pi \pm 2K\pi , \quad x_2 = 0 , \quad K = 1,2,\ldots \qquad (3.56)$$

The position $x_1 = x_2 = 0$ has been shown to be asymptotically stable in probability. As the system (3.43) is periodic in x_1 , the stability of the origin implies the stability of all positions (3.55). By mechanical and topological reasons the interjacent positions

(3.56) must be unstable. The instability of the positions (3.56) is confirmed by an inspection of $W(x)$. The conditions $W(0) = 0$, $W(x) > 0$ for $x \neq 0$, required by the theorem 2.4, hold only for $|x| < \pi$! These considerations are directly connected with the conception of the stability regions. For the case of eq. (3.43) see [15].

Remark 2 : In this textbook the Liapunov function $v(x)$ was constructed for the case $g(x_1) = \sin x_1$. A similar Liapunov function for an arbitrary $g(x)$ satisfying conditions (3.16) and (3.17) is given in [15].

3. 5. 3 Discussion

1. By means of the stability theorems of section 2 it can be shown, that the stability of the second moments of the linearized eq. (3.28) implies the stability in probability of the original nonlinear eq. (3.27). Thus, the equilibrium position of the pitch motion (3.27) is stable in probability for every function $g(x)$ with the properties (3.16), (3.17), if condition (3.38) is satisfied, where μ is given by (3.26).

2. Since $\mu = 1$ for $g(x_1) = \sin x_1$, the condition (3.54) obtained by Liapunov techniques seems to be a special case of (3.38). Both methods -

covariance matrix and Liapunov function - seem to lead to the same result. This conclusion is erroneous: linearizing and using covariance matrix we state the stability of the equilibrium position for sufficient small initial deviations. The domain of attraction of the origin remains completely unknown. Using Liapunov techniques this domain can be estimated, see [3], [15] and the remark 1 in 3.5.2.

3. The stability conditions (3.38) and (3.54) are given explicitely as conditions on δ_0. For technical applications it is convenient to look at δ_0 as at a given quantity and to rewrite the condition (3.38) (or (3.54)) as

$$\ell < \frac{1}{2}\left(1 - \frac{\delta_0^2}{2\eta_0 - \delta_0^2\eta_0^2}\right). \qquad (3.57)$$

Now the stability domains can be given in the (ℓ,η_0)-plane, i.e. in the plane "gravity gradient torque versus dynamic coefficient of the aerodynamic torque".

4. In the case of a stochastically undisturbed atmosphere pitch motion is stable for

$$1 - 2\ell > 0 .$$

For an atmosphere disturbed by white noise the sharper

condition (3.57) or

$$1 - 2\iota > \frac{\delta_0^2}{2\eta_0 - \delta_0^2\eta_0^2} > 0$$

has to be fulfilled. Thus, accounting atmospheric disturbances the bounds for the destabilizing gravity gradient torque become smaller.

4. Dispersions of the State Variables During Re-Entry

The uncertainties of the atmosphere cause deviations from the precalculated re-entry flight path data. Modelling the density by a diffusion process the vehicle's altitude h, arc length s, velocity v and flight pass angle γ referenced to the local horizon (Fig. 4.1) become stochastic processes too.

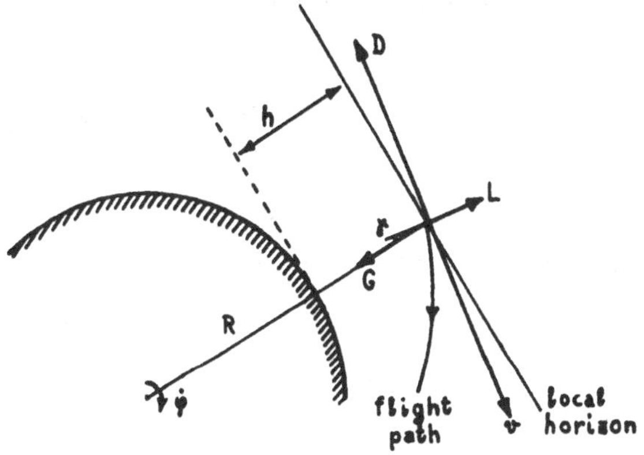

Fig. 4.1.
To the re-entry equations.

Supposing the density to be normally distributed the deviations of the flight path data h, s, v, and γ can be described by their means and variances and expressed in terms of density's mean and variance. These last quantities can be predetermined independently of the re-entry.

Various stochastic models of the density are possible. In this section a simple model without shaping filter is used.

4.1 Re-Entry Equations [18],[19]

The re-entering vehicle is influenced by the lift L, the drag D, the gravitational force G, and the centrifugal force. The motion is described by the equations (Fig. 4.1) :

$$
\begin{aligned}
m\dot{v} &= -D - G\sin\gamma , \\
0 &= L - G\cos\gamma + mv(\dot{\varphi} - \dot{\gamma}) , \\
\dot{h} &= v\sin\gamma , \\
(R + h)\dot{\varphi} &= v\cos\gamma .
\end{aligned}
\tag{4.1}
$$

where m denotes vehicle mass, $G = mg$, and g depends on the altitude : $g = g(h)$. As

$$g(0) = g_0 = \Gamma/R^2$$

and

$$g(h) = \Gamma/(R + h)^2$$

the gravitational force G can be written as

$$G = m\frac{R^2}{(R + h)^2}g_0 . \tag{4.2}$$

Now eqs. (4.1) become

$$\begin{aligned}
\dot{v} &= -\frac{D}{m} - g_0 \frac{R^2}{(R+h)^2} \sin\gamma \,, \\
v(\dot{\gamma} - \dot{\varphi}) &= \frac{L}{m} - g_0 \frac{R^2}{(R+h)^2} \cos\gamma \,, \\
\dot{h} &= v \sin\gamma \,, \\
\dot{\varphi} &= \frac{v}{R+h} \cos\gamma \,.
\end{aligned} \tag{4.3}$$

For re-entering vehicles $h \ll R$ and thus $R + h \approx R$. Introducing the arc length $s = R\varphi$ we get the usual re-entry equations

$$\begin{aligned}
\dot{v} &= -\frac{D}{m} - g_0 \sin\gamma \\
\dot{\gamma} &= \frac{L}{mv} - \left(\frac{g_0}{v} - \frac{v}{R}\right) \cos\gamma \\
\dot{h} &= v \sin\gamma \\
\dot{s} &= v \cos\gamma \,.
\end{aligned} \tag{4.4}$$

Drag and lift are given by

$$\begin{aligned}
L &= \frac{1}{2} \varrho v^2 F_0 c_L \\
D &= \frac{1}{2} \varrho v^2 F_0 c_D
\end{aligned} \tag{4.5}$$

where F_0 is the vehicle's reference area and c_L, c_D denote the lift and drag coefficients, respectively.

4.2 The Stochastic Model of the Atmospheric Density

Considering in section 3 a satellite in circular orbit, the density could be modeled by the simple expression $\varrho = \varrho_0 + \dot{w}$, ϱ_0 const. During re-entry the density changes with altitude and is given (in the deterministic case) by

$$(4.6)\qquad \varrho(h) = \varrho(0)c^{-\frac{h}{d}} .$$

The quantity d depends on the altitude h and can be considered as a constant only for bounded intervals of the altitude. However, during re-entry some constant mean value can be taken for d. Then the density satisfies the deterministic differential equation

$$(4.7)\qquad \frac{d\varrho}{dh} = -\frac{1}{d}\varrho .$$

The various fluctuations of the density [20], [21] can be accounted by the corresponding stochastic model

$$(4.8)\qquad d\varrho(h) = -\frac{1}{d}\varrho(h)dh + \delta dw(h) .$$

The density is now composed of a deterministic part ϱ_r satisfying (4.7), and a stochastic process $\Delta\varrho$

$$(4.9)\qquad \varrho = \varrho_r + \Delta\varrho .$$

Inserting eq. (4.9) into eq. (4.8) and regarding eq. (4.7) we obtain the stochastic differential equation describing the fluctuations $\Delta\varrho$:

$$d\Delta\varrho = -\frac{1}{d}\Delta\varrho dh + \delta dw(h) . \qquad (4.10)$$

The solution of eq. (4.10) is given by formula (2.44):

$$\Delta\varrho(h) = e^{-\frac{h}{d}}\left[\Delta\varrho(h) + \delta\int_0^h e^{+\frac{h}{d}} dw(h)\right] . \qquad (4.11)$$

Supposing $\Delta\varrho(0)$ not to be correlated with $w(h)$, the mean value and the covariance of the density fluctuations are

$$E\Delta\varrho(h) = e^{-\frac{h}{d}} E\Delta\varrho(0) \qquad (4.12)$$

and

$$E\left[\Delta\varrho(h)\right]^2 = \sigma^2(h) = e^{-\frac{2h}{d}}\left[E\left[\Delta\varrho(0)\right]^2 + \delta^2\frac{d}{2}(e^{\frac{2h}{d}} - 1)\right] =$$
$$= \left[E\left[\Delta\varrho(0)\right]^2 - \frac{d}{2}\delta^2\right]e^{-\frac{2h}{d}} + \frac{d}{2}\delta^2 . \qquad (4.13)$$

It is naturally to suppose $E\Delta\varrho(0) = 0$. Then

$$E\Delta\varrho(h) = 0 \qquad (4.12')$$

and

$$E[\Delta\varrho(h)]^2 = \sigma^2(h) = \frac{d}{2}\delta^2(1 - e^{-\frac{2h}{d}}) . \tag{4.13'}$$

The cross correlation of fluctuations in different altitudes is

$$R(h,h + \Delta h) = E\Delta\varrho(h)\Delta\varrho(h + \Delta) =$$

$$= \left(E[\Delta\varrho(0)]^2 - \frac{d}{2}\delta^2\right)e^{-\frac{2h+|\Delta h|}{d}} + \frac{d}{2}\delta^2 e^{-\frac{|\Delta h|}{d}}$$

or using eq. (4.13)

$$R(h,h + \Delta h) = \sigma^2(h)e^{-\frac{|\Delta h|}{d}} . \tag{4.14}$$

The obtained variance (4.13) and cross correlation function (4.14) fit our intuition and our observations [21] : the variance increases with altitude, the cross correlation decreases with altitude difference (Fig. 4.2).

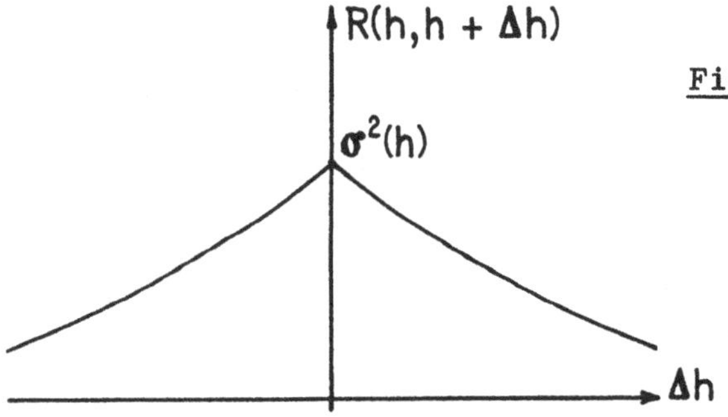

Fig. 4.2.

The cross-correlation of the fluctuations of the atmospheric density.

For high altitudes

$$\sigma^2(h) = \frac{d}{2}\delta^2$$

and thus, the white noise intensity can be expressed approximately by the variance

$$\delta^2 = \frac{2\sigma^2(h)}{d}. \tag{4.15}$$

4.3 Fluctuations Caused by Density

The fluctuations of the density cause fluctuations of the state variables s, v, γ and h. Denoting these fluctuations by Δs, Δv, $\Delta\gamma$, Δh, we have

$$s = s_r + \Delta s\,,\ v = v_r + \Delta v\,,\ \gamma = \gamma_r + \Delta\gamma\,,\ h = h_r + \Delta h\,. \tag{4.16}$$

Here, the quantities s_r, v_r, γ_r, h_r are deterministic and correspond to the mean value ϱ_r of the density, whereas Δs, Δv, $\Delta\gamma$, Δh are stochastic and correspond to the fluctuation $\Delta\varrho$.

Supposing all deviations to be small, inserting the relations (4.5) and (4.9) into the eqs. (4.4), and linearizing with respect to the Δ-terms, we get a system describing the processes Δs, Δv, $\Delta\gamma$ and Δh:

$$\Delta\dot{V} = -\frac{\varrho_r v_r F_0 c_D}{m}\Delta V - g_0\cos\gamma_r\,\Delta\gamma - \frac{v_r^2 F_0 c_D}{2m}\Delta\varrho$$

$$\Delta\dot{\gamma} = \left[\frac{\varrho_r F_0 c_L}{2m} + \left(\frac{g_0}{v_r^2} + \frac{1}{R}\right)\cos\gamma_r\right]\Delta V + \left(\frac{g_0}{v_r} - \frac{v_r}{R}\right)\sin\gamma_r\,\Delta\gamma + \frac{v_r F_0 c_L}{2m}\Delta\varrho$$

(4.17)

$$\Delta\dot{h} = \sin\gamma_r\,\Delta V + v_r\cos\gamma_r\,\Delta\gamma$$

$$\Delta\dot{s} = \cos\gamma_r\,\Delta V - v_r\sin\gamma_r\,\Delta\gamma\ .$$

Since the fluctuation $\Delta\varrho$ of the density depends on the altitude, it is convenient to introduce the reference altitude h_r as the new independent variable. Dividing for this purpose eqs. (4.17) by equation $\dot{h}_r = v_r \sin\gamma_r$ we obtain

$$\frac{d\Delta v}{dh_r} = -\frac{\varrho_r F_0 c_D}{m\sin\gamma_r}\Delta V - \frac{g_0}{v_r}\cot\gamma_r\,\Delta\gamma - \frac{F_0 c_D v_r}{2m\sin\gamma_r}\Delta\varrho\ ,$$

$$\frac{d\Delta\gamma}{dh_r} = \left[\frac{\varrho_r F_0 c_L}{2m v_r\sin\gamma_r} + \left(\frac{g_0}{v_r^2} + \frac{1}{R}\right)\frac{\cot\gamma_r}{v_r}\right]\Delta V +$$

(4.18)

$$+\frac{1}{v_r}\left(\frac{g_0}{v_r} - \frac{v_r}{R}\right)\Delta\gamma + \frac{F_0 c_L}{2m\sin\gamma_r}\Delta\varrho\ ,$$

$$\frac{d\Delta h}{dh_r} = \frac{1}{v_r}\Delta V + \cot\gamma_r\,\Delta\gamma\ ,$$

$$\frac{d\Delta s}{dh_r} = \frac{\cot\gamma_r}{v_r}\Delta V - \Delta\gamma\ .$$

Completing this system by eq. (4.10) we get finally the stochastic system

$$dx(h_r) = Ax(h_r)dh_r + Gdw(h_r) \tag{4.19}$$

with the state vector

$$x = \begin{bmatrix} \Delta v \\ \Delta \gamma \\ \Delta h \\ \Delta s \\ \Delta \varrho \end{bmatrix}$$

and the matrices

$$G = \begin{bmatrix} 0 \\ 0 \\ 0 \\ 0 \\ \delta \end{bmatrix}$$

and A as given on the next page .

4. 4 Simplified Models

Several authors adapt the assumption

$$\frac{D}{m} \gg g_0 \sin\gamma ,$$

which certainly holds at least in the lower part of the atmosphere. Furthermore, as γ is small, $\dot{s} = v$ can be supposed. Finally, discussing the dispersions of the state variables, we are mainly interested in s and v.

$$
A = \begin{bmatrix}
-\dfrac{\varrho_r F_0 c_D}{m \sin\gamma_r} & -\dfrac{g_0}{v_r}\cot\gamma_r & 0 & 0 & -\dfrac{F_0 v_r c_D}{2m \sin\gamma_r} \\[2ex]
\dfrac{\varrho_r F_0 c_L}{2m v_r \sin\gamma_r} + \left(\dfrac{g_0}{v_r^2} + \dfrac{1}{R}\right)\dfrac{\cot\gamma_r}{v_r} & \dfrac{1}{v_r}\left(\dfrac{g_0}{v_r} - \dfrac{v_r}{R}\right) & 0 & 0 & \dfrac{F_0 c_L}{2m \sin\gamma_r} \\[2ex]
\dfrac{1}{v_r} & \cot\gamma_r & 0 & 0 & 0 \\[2ex]
\dfrac{\cot\gamma_r}{v_r} & -1 & 0 & 0 & 0 \\[2ex]
0 & 0 & 0 & 0 & -\dfrac{1}{d}
\end{bmatrix}
$$

Thus, eqs. (4.4) can be replaced by

$$\dot{v} = -\frac{D}{m}$$

$$\dot{h} = v \sin\gamma \tag{4.20}$$

$$\dot{s} = v .$$

The corresponding equations for the deviations are

$$\Delta\dot{v} = -\frac{F_0 c_D v_r \varrho_r}{m}\Delta v - \frac{F_0 c_D v_r^2}{2m}\Delta\varrho$$

$$\Delta\dot{s} = \Delta v .$$

Dividing by $\dot{h}_r = v_r \sin\gamma_r$ and completing by the equation for the fluctuation of the density :

$$\frac{d\Delta v}{dh_r} = -\frac{F_0 c_D \varrho_r}{m \sin\gamma_r}\Delta v - \frac{F_0 c_D v_r}{2m \sin\gamma_r}\Delta\varrho$$

$$\frac{d\Delta s}{dh_r} = \frac{1}{v_r \sin\gamma_r}\Delta v \tag{4.21}$$

$$\frac{d\Delta\varrho}{dh_r} = -\frac{1}{d}\Delta\varrho + \delta\dot{w} .$$

Rewriting this as a stochastic Itô system, we obtain

$$dx(h_r) = A_1 x(h_r) dh_r + G_1 dw(h_r) \tag{4.22}$$

with

$$x = \begin{bmatrix} \Delta v \\ \Delta s \\ \Delta \varrho \end{bmatrix},$$

$$A_1 = \begin{bmatrix} -\dfrac{F_0 c_D \varrho_r}{m \sin\gamma_r} & 0 & -\dfrac{F_0 c_D v_r}{2m \sin\gamma_r} \\ \dfrac{1}{v_r \sin\gamma_r} & 0 & 0 \\ 0 & 0 & -\dfrac{1}{d} \end{bmatrix},$$

$$G_1 = \begin{bmatrix} 0 \\ 0 \\ \delta \end{bmatrix}.$$

A further simplification is supposed to be possible in [22]. The author assumes

$$\frac{F_0 c_D \varrho_r}{m \sin\gamma_r} \Delta v \ll \frac{F_0 c_D v_r}{2m \sin\gamma_r} \Delta \varrho$$

and neglects the first term on the right side of the first eq. (4.21). However, this assumption is equivalent to

$$\frac{\Delta v}{v_r} \ll \frac{1}{2} \frac{\Delta \varrho}{\varrho_r}$$

and seems not to be justified a priori.

4.5 Covariance Equation

The covariance equations for the systems (4.19) or (4.22) are given by eq. (2.56). As the covariance equation of the system (4.19) is equivalent to a linear system of the order of 15, only the linear system corresponding to the covariance equation of the simplified model (4.22) may be given explicitly. With the abbreviations

$$a_{11} = -\frac{F_0 c_D \varrho_r}{m \sin\gamma_r}, \qquad a_{13} = -\frac{F_0 c_D v_r}{2m \sin\gamma_r},$$

$$a_{21} = -\frac{1}{v_r \sin\gamma_r}, \qquad a_{33} = -\frac{1}{d},$$

and

$$p_{11} = E(\Delta v)^2, \quad p_{12} = E\Delta v\, \Delta s, \quad p_{13} = E\Delta v\, \Delta\varrho$$

$$p_{22} = E(\Delta s)^2, \quad p_{23} = E\Delta s\, \Delta\varrho, \quad p_{33} = E(\Delta\varrho)^2,$$

this system is given by

$$\begin{aligned} \dot{p}_{11} &= 2a_{11}p_{11} + 2a_{13}p_{13} \\ \dot{p}_{12} &= a_{21}p_{11} + a_{11}p_{12} + a_{13}p_{23} \\ \dot{p}_{13} &= (a_{11} + a_{33})p_{13} + a_{13}p_{33} \end{aligned} \qquad (4.23a)$$

$$\dot{p}_{22} = 2a_{21}p_{12}$$

$$\dot{p}_{23} = a_{21}p_{13} + a_{33}p_{23} \tag{4.23b}$$

$$\dot{p}_{33} = 2a_{33}p_{33} + \delta^2 .$$

As the coefficients a_{11} , a_{13} , a_{21} depend on the variable quantities ϱ_r, v_r , γ_r the system (4.23) must be solved simultaneously with the systems for ϱ_r , v_r , γ_r :

$$\frac{dv_r}{dh_r} = -\frac{\varrho_r v_r F_0 c_D}{2m \sin\gamma_r},$$

$$\frac{d\gamma_r}{dh_r} = \frac{\varrho_r F_0 c_L}{2m \sin\gamma_r} - \left(\frac{g_0}{v_r^2} - \frac{1}{R}\right)\cot\gamma_r , \tag{4.24}$$

$$\frac{d\varrho_r}{dh_r} = -\frac{1}{d}\varrho_r .$$

4.6 Dispersions of Arc Length and Velocity

The described stochastic method for the determination of the dispersions of the flight path data during re-entry was first proposed in [22]. The author uses a simplified stochastic model and re- stricts the investigation to the dispersions of arc length s and velocity v ; in [22] some numerical results are given too.

5. Stabilization by the Magnetic Field of the Earth

The magnetic field of the earth is approximately a dipol field (fig. 5.1). The dipol axis is inclined to the earth's axis, but the inclination is usually neglected. The stabilization by means of the earth's magnetic field is based on the interaction between this field and the magnetic field of the satellite. However, an ideal stabilization is possible only for satellites in equatorial orbits [12].

5.1 Satellite in Equatorial Orbit

Consider the plane yaw oscillations of a satellite in a circular and equatorial orbit (Fig. 5.2 and fig. 3.1 ; see pages 93 and 48).

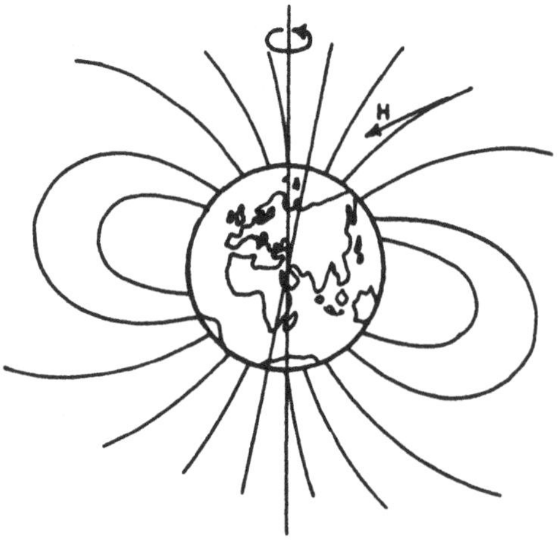

Fig. 5.1.

The magnetic field of the earth.

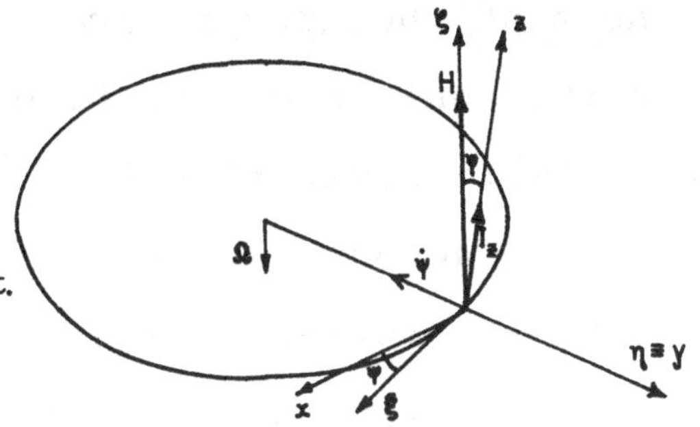

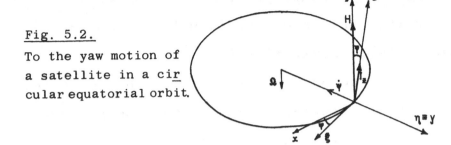

Fig. 5.2.
To the yaw motion of a satellite in a circular equatorial orbit.

The total intensity H of the earth's field is orthogonal to the orbital plane. Suppose, the total magnetic moment of the satellite is $I = (0, 0, I_z) = \text{const}$. The moment can be caused for instance by a magnetic bar in the satellite-fixed z-axis. Then there is a restoring torque

$$M_y^I = -HI_z \sin\Psi \tag{5.1}$$

exerted by the earth's field on the satellite.

A second torque is due to the eddy currents in the shell of the satellite. These currents arise when the satellite is moving in the earth's magnetic field, and they cause by intersection with the field of the earth the damping torque

$$M_y^{II} = -KH^2\dot{\Psi} . \tag{5.2}$$

The factor $K = \text{const.}$ in eq. (5.2) depends on the thickness and the electromagnetic properties of the shell and on the moments of inertia of the satellite.

The y-component of the gravity gradient torque is zero even in the nonlinear case (see (3.6) and (3.7)) if the satellite is symmetric ($A = C$).

The aerodynamic torques can be neglected if the orbit is sufficiently high. Thus, the yaw oscillations of the satellite are described by

$$(5.3) \qquad B\ddot{\Psi} = -KH^2\dot{\Psi} - HI_z \sin\Psi .$$

Remark : For further information on magnetic stabilization see [12].

5.2 Fluctuations of the Magnetic Field of the Earth [20] and the Stochastic Model of Yaw Motion

If the quantities B, H, K, I_z in eq. (5.3) are constants, the yaw motion $\Psi = \Psi(t)$ is asymptotically stable. However, the assumption H = const is not justified in reality. The total intensity H of the earth's field is always corrupted by noise :

$$H = H_0 + \Delta H$$

and supposing the noise ΔH to be small

$$H^2 = H_0^2 + 2H_0 \Delta H .$$

The fluctuation of H depends on the activity of the sun and have slowly and rapidly varying frequencies and amplitudes and can be modeled by the white noise

$$\Delta H = H_0 \delta \dot{w} \qquad (5.4)$$

where δ is the intensity of the noise and $\dot{w}$ denotes the white noise.

Replacing H and H^2 in eq. (5.3) by the stochastic models

$$H = H_0(1 + \delta \dot{w}) \qquad (5.5)$$

and

$$H^2 = H_0^2(1 + 2\delta \dot{w}) \qquad (5.6)$$

and separating deterministic and stochastic terms, the stochastic system describing the yaw oscillations is obtained

$$d\begin{bmatrix}\Psi \\ \omega\end{bmatrix} = \begin{bmatrix}\omega \\ -\frac{H_0 I_z}{B}\sin\Psi - \frac{KH_0^2}{B}\omega\end{bmatrix}dt + \begin{bmatrix}0 \\ -\frac{H_0 I_z \delta}{B}\sin\Psi - \frac{2KH_0^2\delta}{B}\omega\end{bmatrix}dw. \qquad (5.7)$$

5.3 Stability of the yaw Motion

Introducing for convenience the new constants

$$\alpha = \frac{H_0 I_z}{B} > 0, \quad \beta = \frac{KH_0^2}{B} > 0 \qquad (5.8)$$

eq. (5.7) can be rewritten as

$$(5.9)\qquad d\begin{bmatrix}\Psi\\ \omega\end{bmatrix} = \begin{bmatrix}\omega\\ -\alpha\sin\Psi - \beta\omega\end{bmatrix}dt + \begin{bmatrix}0\\ -\alpha\delta\sin\Psi - 2\beta\delta\omega\end{bmatrix}dw\,.$$

The linearized equation is

$$(5.10)\qquad d\begin{bmatrix}\Psi\\ \omega\end{bmatrix} = \begin{bmatrix}0 & 1\\ -\alpha & -\beta\end{bmatrix}\begin{bmatrix}\Psi\\ \omega\end{bmatrix}dt + \begin{bmatrix}0 & 0\\ -\alpha\delta & -2\beta\delta\end{bmatrix}\begin{bmatrix}\Psi\\ \omega\end{bmatrix}dw\,.$$

5. 3. 1 Stability of the Second Moments

The investigation of the stability of the second moments $E\Psi^2$, $E\Psi\omega$, $E\omega^2$ for the linearized eq. (5.10) runs exactly as in section 3.5.1 and will not be repeated here. The stability condition for the second moments is

$$(5.11)\qquad \delta^2 < \frac{2\beta}{\alpha + 4\beta^2}\,,\quad \alpha > 0\,,\quad \beta > 0\,.$$

5. 3. 2 Stability of the Nonlinear Equation

As Liapunov function will serve

$$(5.12)\qquad V = a\Psi^2 + b\Psi\omega + c\omega^2 + f\sin^2\frac{\Psi}{2}\,.$$

The function V is positive definite for

$$(5.13)\qquad a > 0$$

$$4ac > b^2 \tag{5.14}$$

$$f > 0 . \tag{5.15}$$

Applying the L-operator (see sections 2.7.6 to 2.7.8) we obtain

$$\begin{aligned} LV &= (2a\Psi + b\omega + \tfrac{1}{2}f\sin\Psi)\omega + \\ &\quad + (b\Psi + 2c\omega)(-\alpha\sin\Psi - \beta\omega) + \\ &\quad + c(-\alpha\delta\sin\Psi - 2\beta\delta\omega)^2 \\ &= (2a - b\beta)\Psi\omega + (\tfrac{1}{2}f - 2c\alpha + 4c\delta^2\alpha\beta)\omega\sin\Psi - \\ &\quad - (2c\beta - b - 4c\beta^2\delta^2)\omega^2 - b\alpha\Psi\sin\Psi + \\ &\quad + c\alpha^2\delta^2\sin^2\Psi . \end{aligned} \tag{5.16}$$

As the mixed terms $\Psi\omega$ and $\omega\sin\Psi$ in eq. (5.16) have to vanish, the constants must satisfy the conditions

$$a = \frac{1}{2}b\beta \tag{5.17}$$

$$f = 4c\alpha(1 - 2\beta\delta^2) . \tag{5.18}$$

It is $a > 0$, $\beta > 0$ due to eqs. (5.13) and (5.8), respectively. Thus $b > 0$ according to eq. (5.17) and we can fix

$$b = 1 . \tag{5.19}$$

In consequence of eqs. (5.15) and (5.18) we must require

$$\delta^2 < \frac{1}{2\beta} . \tag{5.20}$$

Now LV reads

$$LV = -(2c\beta - 1 - 4c\beta^2\delta^2)\omega^2 - \alpha\Psi \sin\Psi(1 - c\alpha\delta^2 \frac{\sin\Psi}{\Psi})$$

and it is

$$LV \leq -(2c\beta - 1 - 4c\beta^2\delta^2)\omega^2 - \alpha(1 - c\alpha\delta^2)\Psi \sin\Psi . \tag{5.21}$$

In order to satisfy the conditions of theorem 2.4, the following inequalities must hold :

$$\delta^2 < \frac{2c\beta - 1}{4c\beta^2} \tag{5.22}$$

$$\delta^2 < \frac{1}{c\alpha} . \tag{5.23}$$

As $(2c\beta - 1)/4c\beta^2 < 1/2\beta$, the condition (5.20) drops out.

It remains to fix the constant c. As in section 3.5.2 the best c is obtained requiring

$$\frac{2c\beta - 1}{4c\beta^2} = \frac{1}{c\alpha} .$$

This gives

$$c = \frac{\alpha + 4\beta^2}{2\alpha\beta} . \tag{5.24}$$

Now the functions V and LV can be written as

$$\left.\begin{aligned} V &= \frac{1}{2}\left(\sqrt{\beta}\,\Psi + \frac{\omega}{\sqrt{\beta}}\right)^2 + \frac{2\beta}{\alpha}\omega^2 + \\ &+ 4(\alpha + 4\beta^2)\left(\frac{1}{2\beta} - \delta^2\right)\sin^2\frac{\Psi}{2}\,, \end{aligned}\right\} \qquad (5.27)$$

$$LV \leq -\frac{\alpha + 4\beta^2}{2\alpha\beta}\left(\frac{2\beta}{\alpha + 4\beta^2} - \delta^2\right)(\alpha^2\Psi\sin\Psi + 4\beta^2\omega^2)\,. \qquad (5.26)$$

All conditions of theorem 2.4 are fulfilled for

$$\delta^2 < \frac{2\beta}{\alpha + 4\beta^2}\,, \quad \alpha > 0\,, \quad \beta > 0\,, \qquad (5.27)$$

i.e. the nonlinear yaw oscillation (5.9) is asymptotically stable in probability, if the noise intensity satisfies condition (5.27).

Returning to the original constants (5.8) the stability conditions read

$$\delta^2 < \frac{2KH_0B}{I_zB + 4K^2H_0^3}\,, \quad K > 0\,, \quad I_z > 0\,. \qquad (5.27')$$

5.3.3 Discussion

1. The term $(\dot{w})^2$ in eq. (5.6) was neglected to avoid mathematical difficulties. In consequence the damping of the models (5.9) and (5.10)

is smaller than the damping of the real system (5.3). Therefore the stability domain given by condition (5.11) or (5.27) is smaller than the stability domain of the real system (5.3).

2. As in section 3 the conditions (5.11) for the stability in mean square and (5.27) for the stability in probability turn out to be identical. That is by no means a general property of stochastic systems. The domain of stability in mean square is fixed for a given system. The domain of stability in probability depends on the choice of the Liapunov function. For a bad Liapunov function the domain can be smaller, for a good Liapunov function the domain can be larger than the domain of the stability in mean square. For examples see [3], [5] and the section 3.5.2.

6. Satellites with Moving Parts

In this section two completely independent examples of satellites with moving parts are considered. In both cases the noise is caused by the non-rigidity of the satellite. However, the resulting stochastic models are quite different. In the first case the mass distribution of the satellite remains constant in spite of the moving parts. The problem leads to a common linear stochastic differential equation. In the second case the mass distribution of the satellite varies in dependence of the inner motion. The stochastic modeling results in an unusual nonlinear stochastic differential equation.

6.1 Stabilization by Wheels with Noisy Frictional Torque

Actively stabilizing the orientation of a satellite, various devices can be used to generate the stabilizing control torques. Here, stabilization by fly-wheels will be investigated. For simplicity reasons only a very rough theory is given.

6.1.1 The Deterministic Model

Consider a satellite with three wheels rotating around the principal axes of inertia of the system satellite + wheels (Fig. 6.1). The principal moments of inertia of the system are A, B, C. The axial moments of inertia of the wheels are $I_1 = I_2 = I_3 = I$. The absolute angular velocity of the satellite is $\bar{\omega} = (\omega_x, \omega_y, \omega_z)$. The angular velocities of the wheels with respect to the satellite are ν_1, ν_2, ν_3.

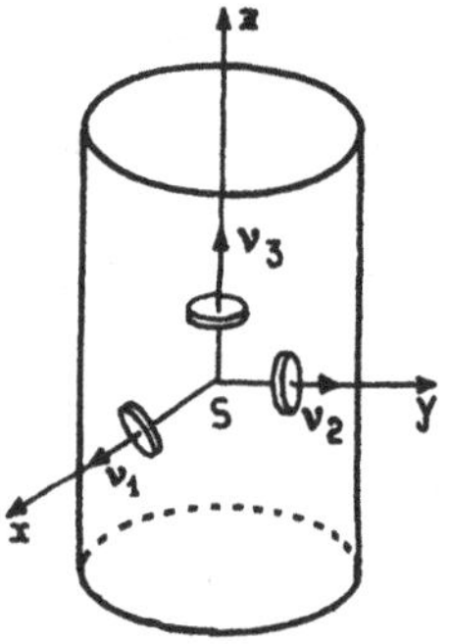

Fig. 6.1.

Satellite with fly-wheels.

The moment of momentum $\bar{D}$ of the system satellite + wheels has the components

$$
\begin{aligned}
D_x &= A\omega_x + I\nu_1 \\
D_y &= B\omega_y + I\nu_2 \qquad (6.1) \\
D_z &= C\omega_z + I\nu_3 .
\end{aligned}
$$

Inserting these components into the Euler equation

$$\frac{d\bar{D}}{dt} + \bar{\omega} \times \bar{D} = \bar{M}_{ex}$$

the equations of motion of the satellite are obtained

$$\begin{aligned} A\dot{\omega}_x + (C - B)\omega_y\omega_z &= I(\nu_2\omega_z - \nu_3\omega_y - \dot{\nu}_1) + M_{xex} \\ B\dot{\omega}_y + (A - C)\omega_z\omega_x &= I(\nu_3\omega_x - \nu_1\omega_z - \dot{\nu}_2) + M_{yex} \qquad (6.2) \\ C\dot{\omega}_z + (B - A)\omega_x\omega_y &= I(\nu_1\omega_y - \nu_2\omega_x - \dot{\nu}_3) + M_{zex} . \end{aligned}$$

The motion of the wheels is described by

$$\begin{aligned} I_1(\dot{\omega}_x + \dot{\nu}_1) &= u_1 - k_1\nu_1 \\ I_2(\dot{\omega}_y + \dot{\nu}_2) &= u_2 - k_2\nu_2 \qquad I_i = I, \quad K_i = K \qquad i = 1,2,3 . \qquad (6.3) \\ I_3(\dot{\omega}_z + \dot{\nu}_3) &= u_3 - k_3\nu_3 \end{aligned}$$

Here, u_1, u_2, u_3 denote the driving moments and $k\nu_1$, $k\nu_2$, $k\nu_3$ the frictional moments, respectively. The coefficient k depends on various factors as temperature, angular velocity etc., and will be considered later on as a noisy quantity.

Suppose, the prescribed orientation of the satellite is given by the inertial axes x_0, y_0, z_0 - The position of the x, y, z-axes with respect to the x_0, y_0, z_0-axes can be described by the Euler angles φ, Ψ, ϑ . Supposing the deviations φ, Ψ, ϑ and the velocities $\dot{\varphi}, \dot{\Psi}, \dot{\vartheta}$ to be small, the axes transform according to the table

(6.4)

	x_0	y_0	z_0
x	1	ϑ	$-\psi$
y	$-\vartheta$	1	φ
z	ψ	$-\varphi$	1 .

The linearized angular velocities are

$$\omega_x = \dot{\varphi}\,, \quad \omega_y = \dot{\psi}\,, \quad \omega_z = \dot{\vartheta}\,. \tag{6.5}$$

Using (6.5) and linearizing eqs. (6.2) and (6.3) we obtain

$$\begin{cases} A\ddot{\varphi} = I(\nu_2\dot{\vartheta} - \nu_3\dot{\psi} - \dot{\nu}_1) + M_{x\,ex} \\ B\ddot{\psi} = I(\nu_3\dot{\varphi} - \nu_1\dot{\vartheta} - \dot{\nu}_2) + M_{y\,ex} \\ C\ddot{\vartheta} = I(\nu_1\dot{\psi} - \nu_2\dot{\varphi} - \dot{\nu}_3) + M_{z\,ex} \end{cases} \tag{6.6}$$

$$\begin{cases} I(\ddot{\varphi} + \dot{\nu}_1) = u_1 - k\nu_1 \\ I(\ddot{\psi} + \dot{\nu}_2) = u_2 - k\nu_2 \\ I(\ddot{\vartheta} + \dot{\nu}_3) = u_3 - k\nu_3\,. \end{cases} \tag{6.7}$$

In order to stabilize the satellite's axes x, y, z in the prescribed position x_0, y_0, z_0 (i.e. to make the deviations φ, $\dot{\varphi}$, Ψ, $\dot{\Psi}$, ϑ, $\dot{\vartheta}$ vanish), the driving torques must be properly chosen functions of the deviations. To get the direct dependence of the driving torques u_1, u_2, u_3 on the deviations φ, Ψ, . . . , $\dot{\vartheta}$ the quantities $v_1, v_2, v_3, \dot{v}_1, \dot{v}_2, \dot{v}_3$ in eqs. (6.6) and (6.7) must be eliminated first. As the usual elimination procedure is very cumbersome, a sophisticated method [23] will be used in slightly simplified form. Denote by h_1 , h_2 , h_3 the components of $\bar{D}$ on the inertial axes x_0, y_0, z_0. Regarding the transformation scheme (6.4) and using the linearized eq. (6.1) these components are obtained as

$$\begin{aligned} h_1 &= (A\dot{\varphi} + Iv_1)1 + (B\dot{\Psi} + Iv_2)\vartheta + (C\dot{\vartheta} + Iv_3)\Psi \\ h_2 &= (A\dot{\varphi} + Iv_1)\vartheta + (B\dot{\Psi} + Iv_2)1 - (C\dot{\vartheta} + Iv_3)\varphi \qquad (6.8) \\ h_3 &= -(A\dot{\varphi} + Iv_1)\Psi + (B\dot{\Psi} + Iv_2)\varphi + (C\dot{\vartheta} + Iv_3)1 . \end{aligned}$$

As the external torques M_{ex} are very small compared with the artificial inner torques u_1, u_2, u_3, the components h_1, h_2, h_3 can be regarded as constants at least during the reorientation process $\varphi \to 0$, $\Psi \to 0$, $\vartheta \to 0$. Resolving eq. (6.8) with respect to the brackets we obtain :

$$A\dot{\varphi} + Iv_1 = h_1 + h_2\vartheta - h_3\Psi$$

$$
\begin{aligned}
B\dot{\psi} + I\nu_2 &= -h_1\vartheta + h_2 + h_3\varphi \\
C\dot{\vartheta} + I\nu_3 &= h_1\psi - h_2\varphi + h_3 \, .
\end{aligned}
\tag{6.9}
$$

Now $\nu_1, \nu_2, \nu_3, \dot{\nu}_1, \dot{\nu}_2, \dot{\nu}_3$ can be eliminated rapidly from eqs. (6.7) and (6.9) :

$$
\begin{aligned}
&\ddot{\varphi} + \frac{A}{A-I}\frac{K}{I}\dot{\varphi} + \frac{h_3}{A-I}\left(\dot{\psi} + \frac{K}{I}\psi\right) - \frac{h_2}{A-I}\left(\dot{\vartheta} + \frac{K}{I}\vartheta\right) = \\
&\qquad = \frac{h_1}{A-I}\frac{K}{I} - \frac{u_1}{A-I} \, , \\
&\ddot{\psi} + \frac{B}{B-I}\frac{K}{I}\dot{\psi} + \frac{h_1}{B-I}\left(\dot{\vartheta} + \frac{K}{I}\vartheta\right) - \frac{h_3}{B-I}\left(\dot{\varphi} + \frac{K}{I}\varphi\right) = \\
&\qquad = \frac{h_2}{B-I}\frac{K}{I} - \frac{u_2}{B-I} \, , \\
&\ddot{\vartheta} + \frac{C}{C-I}\frac{K}{I}\dot{\vartheta} + \frac{h_2}{C-I}\left(\dot{\varphi} + \frac{K}{I}\varphi\right) - \frac{h_1}{C-I}\left(\dot{\psi} + \frac{K}{I}\psi\right) = \\
&\qquad = \frac{h_3}{C-I}\frac{K}{I} - \frac{u_3}{C-I} \, .
\end{aligned}
\tag{6.10}
$$

This system can be simplified by inspection of the coupling terms. The h_i are components of the moment of momentum $\bar{D}$, i.e. they are products of moments of inertia and angular velocities. Thus, the coupling terms in eqs. (6.10) have the same order of magnitude as the terms $\bar{\omega}\times\bar{D}$ previously neglected linearizing eq. (6.2). Neglecting the coupling terms too, the linearized equations of the system satellite + wheels read as

$$\begin{aligned} \ddot{\varphi} + \frac{A}{A-I}\frac{K}{I}\dot{\varphi} &= \frac{h_1}{A-I}\frac{K}{I} - \frac{u_1}{A-I}, \\ \ddot{\Psi} + \frac{B}{B-I}\frac{K}{I}\dot{\Psi} &= \frac{h_2}{B-I}\frac{K}{I} - \frac{u_2}{B-I}, \\ \ddot{\vartheta} + \frac{C}{C-I}\frac{K}{I}\dot{\vartheta} &= \frac{h_3}{C-I}\frac{K}{I} - \frac{u_3}{C-I}. \end{aligned} \tag{6.11}$$

For $u_i = 0$ the motions (6.11) are obviously unstable. The stabilization is achieved if the driving torques are fixed as follows

$$\begin{aligned} \frac{u_1}{A-I} &= \frac{h_1}{A-I}\frac{K}{I} + a_1\dot{\varphi} + \omega_1^2\varphi \\ \frac{u_2}{B-I} &= \frac{h_2}{B-I}\frac{K}{I} + a_2\dot{\Psi} + \omega_2^2\Psi \\ \frac{u_3}{C-I} &= \frac{h_3}{C-I}\frac{K}{I} + a_3\dot{\vartheta} + \omega_3^2\vartheta . \end{aligned} \tag{6.12}$$

Now the motion of the actively stabilized satellite is described by

$$(6.13)\qquad \begin{aligned} \ddot{\varphi} + \left(\frac{A}{A-I}\frac{K}{I} + a_1\right)\dot{\varphi} + \omega_1^2\varphi &= 0\,, \\ \ddot{\psi} + \left(\frac{B}{B-I}\frac{K}{I} + a_2\right)\dot{\psi} + \omega_2^2\psi &= 0\,, \\ \ddot{\vartheta} + \left(\frac{C}{C-I}\frac{K}{I} + a_3\right)\dot{\vartheta} + \omega_3^2\vartheta &= 0\,. \end{aligned}$$

As $A > I$, $B > I$, $C > I$ and $k > 0$, the satellite is asymptotically stable for

$$(6.14)\qquad a_1 > -\frac{A}{A-I}\frac{K}{I}\,; \quad \omega_1^2 > 0\,. \quad \text{etc.}$$

Fixing $\omega_1^2 = \omega_2^2 = \omega_3^2 = \omega_0^2$ and properly choosing the constants a_i all three librational motions of the stabilized satellite can be described by one equation

$$(6.15)\qquad \ddot{x} + 2\zeta\omega_0\dot{x} + \omega_0^2 = 0\,.$$

6. 1. 2 The Stochastic Model

The assumption that the friction coefficient k is constant is a very rough approximation.

In reality k depends on the angular velocities ν_1, ν_2, ν_3, on temperature, and on other factors. Thus, an exact deterministic model of the friction is very complicated. To discuss the influence of the uncertainties of the friction on the librational motion of the satellite, a stochastic model of the friction can be introduced. Suppose

$$k_i = k_0(1 + \delta_i \dot{w}_i) \tag{6.16}$$

where $k_0 = \text{const}$, $\delta_i = \text{const}$ and $\dot{w}_i$, $i = 1,2,3$, are independent white noises (there is no reason to assume the white noises in the bearings of the three wheels to be equal). The eqs. (6.11) read now as

$$\begin{aligned}
\ddot{\varphi} + \frac{A}{A-I}\frac{K_0}{I}(1 + \delta_1\dot{w}_1)\dot{\varphi} &= \frac{h_1}{A-I}\frac{K_0}{I}(1 + \delta_1\dot{w}_1) - \frac{u_1}{A-I}, \\
\ddot{\psi} + \frac{B}{B-I}\frac{K_0}{I}(1 + \delta_2\dot{w}_2)\dot{\psi} &= \frac{h_2}{B-I}\frac{K_0}{I}(1 + \delta_2\dot{w}_2) - \frac{u_2}{B-I}, \qquad (6.17) \\
\ddot{\vartheta} + \frac{C}{C-I}\frac{K_0}{I}(1 + \delta_3\dot{w}_3)\dot{\vartheta} &= \frac{h_3}{C-I}\frac{K_0}{I}(1 + \delta_3\dot{w}_3) - \frac{u_3}{C-I}.
\end{aligned}$$

Using the same driving torques as above we have

$$\frac{u_1}{A-I} = \frac{h_1}{A-I}\frac{K_0}{I} + a_1\dot{\varphi} + \omega_1^2\varphi$$

$$(6.18)\qquad \begin{aligned} \frac{u_2}{B-I} &= \frac{h_2}{B-I}\frac{K_0}{I} + a_2\dot{\psi} + \omega_2^2\psi \\ \frac{u_3}{C-I} &= \frac{h_3}{C-I}\frac{K_0}{I} + a_3\dot{\vartheta} + \omega_3^2\vartheta \,. \end{aligned}$$

Inserting eqs. (6.18) in eqs. (6.17), the stochastically disturbed equations of the stabilized satellite are obtained :

$$\ddot{\varphi} + \left[\frac{A}{A-I}\frac{K_0}{I}(1 + \delta_1\dot{w}_1) + a_1\right]\dot{\varphi} + \omega_1^2\varphi = \frac{h_1}{A-I}\frac{K_0}{I}\delta_1\dot{w}\,,$$

$$(6.19)\qquad \ddot{\psi} + \left[\frac{B}{B-I}\frac{K_0}{I}(1 + \delta_2\dot{w}_2) + a_2\right]\dot{\psi} + \omega_2^2\psi = \frac{h_2}{B-I}\frac{K_0}{I}\delta_2\dot{w}\,,$$

$$\ddot{\vartheta} + \left[\frac{C}{C-I}\frac{K_0}{I}(1 + \delta_3\dot{w}_3) + a_3\right]\dot{\vartheta} + \omega_3^2\vartheta = \frac{h_3}{C-I}\frac{K_0}{I}\delta_3\dot{w}\,.$$

6.1.3 Stability and Error

As the three equations (6.19) have exactly the same form and are not coupled, it is sufficient to discuss one of them in detail, say the roll equation.

Due to the noise the roll equation is inhomogeneous, and thus an ideal stabilization is impossible. Two questions arise :

1. What are the stability conditions ?

2. What choice of a_1 and ω_1 makes the final error small ?

The discussion of these problems runs as follows : Denoting $\varphi = \xi_1$, and $\dot{\varphi} = \xi_2$ the roll equation can be written as the stochastic system

$$d\begin{bmatrix}\xi_1\\ \xi_2\end{bmatrix} = \begin{bmatrix}0 & 1\\ -\omega_1^2 & -\left(\frac{A}{A-I}\frac{K_0}{I} + a_1\right)\end{bmatrix}\begin{bmatrix}\xi_1\\ \xi_2\end{bmatrix}dt + \left\{\begin{bmatrix}0 & 0\\ 0 & -\frac{A}{A-I}\frac{K_0}{I}\delta_1\end{bmatrix}\begin{bmatrix}\xi_1\\ \xi_2\end{bmatrix} + \begin{bmatrix}0\\ \frac{h_1}{A-I}\frac{K_0}{I}\delta_1\end{bmatrix}\right\}dw_1 . \tag{6.20}$$

The means $E\varphi = E\xi_1 = \bar{\xi}_1$ and $E\dot{\varphi} = E\xi_2 = \bar{\xi}_2$ satisfy the deterministic equation

$$(6.21)\qquad d\begin{bmatrix}\xi_1\\ \xi_2\end{bmatrix} = \begin{bmatrix}0 & 1\\ -\omega_1^2 & -\left(\frac{A}{A-I}\frac{K_0}{I} + a_1\right)\end{bmatrix}\begin{bmatrix}\xi_1\\ \xi_2\end{bmatrix}dt$$

and are asymptotically stable if the conditions

$$(6.22)\qquad \omega_1^2 > 0\ , \quad \frac{A}{A-I}\frac{K_0}{I} + a_1 > 0$$

hold.

To check the stability of the second moments, an extension of formulae (2.51) and (2.56) is convenient. Denote the matrices in eq. (6.20) by $\tilde{A}$, $\tilde{B}$ and $\tilde{C}$. Then eq. (6.20) reads

$$d\xi = \tilde{A}\xi dt + (\tilde{B}\xi + \tilde{C})dw_1$$

and it is

$$d\xi^* = \xi^*\tilde{A}^*dt + (\xi^*\tilde{B}^* + \tilde{C}^*)dw_1 \ .$$

Using Itô formulae it follows as in section 2 :

$$\begin{aligned} d\xi\xi^* &= \xi d\xi^* + (d\xi)\xi^* + (\tilde{B}\xi + \tilde{C})(\xi^*\tilde{B}^* + \tilde{C}^*)dt = \\ &= \xi\xi^*\tilde{A}^*dt + \xi\xi^*\tilde{B}^*dw_1 + \xi\tilde{C}^*dw_1 + \\ &\quad + \tilde{A}\xi\xi^*dt + \tilde{B}\xi\xi^*dw_1 + \tilde{C}\xi^*dw_1 + \end{aligned}$$

$$+ \tilde{B}\xi\xi^*\tilde{B}^*dt + \tilde{B}\xi\tilde{C}^*dt + \tilde{C}\xi^*\tilde{B}^*dt +$$

$$+ \tilde{C}\tilde{C}^*dt \; .$$

Taking the expectation of both sides and regarding section 2.7.2, we obtain for the covariance matrix

$$P = E\xi\xi^*$$

the equation

$$\dot{P} = \tilde{A}P + P\tilde{A}^* + \tilde{B}P\tilde{B}^* + \tilde{C}\tilde{C}^* + \tilde{B}\bar{\xi}\tilde{C}^* + \tilde{C}\bar{\xi}\tilde{B}^*. \quad (6.23)$$

As the mean $\bar{\xi} = E\xi \;\rightarrow\; 0$ with $t \rightarrow \infty$ if condition (6.22) is satisfied, we get finally the result : For $t \rightarrow \infty$ the covariances of eq. (6.20) satisfy the equation

$$\dot{P} = \tilde{A}P + P\tilde{A}^* + \tilde{B}P\tilde{B}^* + \tilde{C}\tilde{C}^* \; . \quad (6.24)$$

This matrix equation is equivalent to the linear nonhomogeneous system

$$\dot{p}_{11} = 2p_{12}$$

$$\dot{p}_{12} = -\omega_1^2 p_{11} - \alpha p_{12} + p_{22} \quad (6.24')$$

$$\dot{p}_{22} = -2\omega_1^2 p_{12} - \beta p_{22} + \gamma^2$$

with

$$\alpha = \frac{A}{A - I}\frac{K_0}{I} + a_1 ,$$

(6.25)
$$\beta = 2\alpha - \left(\frac{A}{A - I}\right)^2 \frac{K_0^2}{I^2}\delta_1^2 ,$$

$$\gamma^2 = \left(\frac{h_1}{A - I}\right)^2 \frac{K_0^2}{I^2}\delta_1^2 .$$

Eliminating p_{12} and p_{22} we obtain the equation for the covariance $p_{11} = E\xi_1^2 = E\varphi^2$:

(6.26) $$\dddot{p}_{11} + (\alpha + \beta)\ddot{p}_{11} + (4\omega_1^2 + \alpha\beta)\dot{p}_{11} + 2\beta\omega_1^2 p_{11} = 2\gamma^2 .$$

The Hurwitz stability conditions for the left side of eq. (6.26) are

$$\alpha + \beta > 0 ,$$

$$4\omega_1^2 + \alpha\beta > 0 ,$$

$$2\beta\omega_1^2 > 0 ,$$

$$(\alpha + \beta)(4\omega_1^2 + \alpha\beta) > 2\beta\omega_1^2 .$$

All conditions are satisfied for $\beta > 0$ i.e. for

(6.27) $$a_1 > \frac{A}{A - I}\frac{K_0}{I}\left[\frac{1}{2}\frac{A}{A - I}\frac{K_0}{I}\delta_1^2 - 1\right] .$$

The stability domains corresponding to the conditions (6.27) and (6.14) are shown in Fig. 6.2.

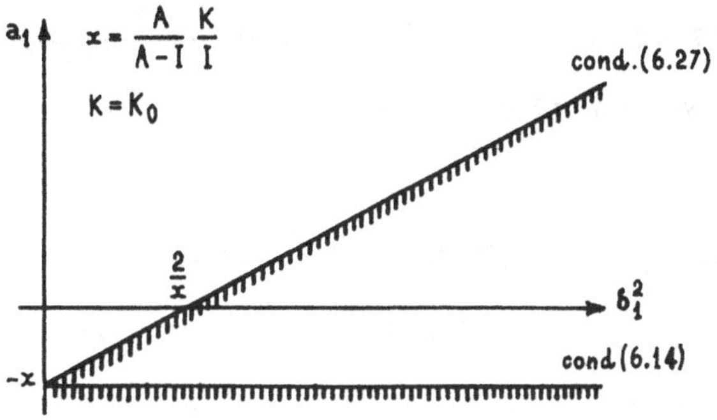

Fig. 6.2.

Stability conditions (6.14) nad (6.27) for the deterministic and stochastic models, respectively.

Comparing condition (6.27) with condition (6.14) for the deterministic system, we see that a sufficiently intensive noise imposes additional requirements on the driving torques (6.12).

If stability condition (6.27) is fulfilled, the covariance p_{11} tends with $t \to \infty$ to the value $\gamma^2/\beta\omega_1^2$ i.e.

$$E\varphi^2 \to \frac{h_1^2 K_0^2 \delta_1^2}{2\omega_1^2 I^2 (A - I)^2} \frac{1}{a_1 - \frac{A}{A - I}\frac{K_0}{I}\left(\frac{1}{2}\frac{A}{A - I}\frac{K_0}{I}\delta_1^2 - 1\right)} \quad (6.28)$$

with $t \to \infty$.

Obviously, for a small final square error the driving coefficients a_1 and ω_1 are to be fixed as large as possible. Thus, an old fact of control engineering is confirmed : the larger the amplification factor the better (as long as stability is not disturbed).

Remark 1 : The influence of the noise expressed by

eqs. (6.27) and (6.28) may often be very small. However, the investigation shows that the satellite with moving parts is influenced by additional torques not known in the case of a completely rigid satellite.

Remark 2 : White noise has been supposed. If sufficient experimental data on the friction are available coloured noise model with shaping filter can be introduced.

6.2 Satellite with Variable Moments of Inertia

As no real satellite is completely rigid the investigation of satellites with variable moments of inertia is very important. During the last fifteen years many attempts have been made in this field and many results were obtained. Nevertheless, a theory rigorous in its statements and convenient for applications is not yet given. Very likely a deterministic version of a theory fitting the requirements mentioned above is not possible at all. Just here, further developed stochastic methods seem to be promising.

In this textbook only the simplest case is sketched. Considered is a satellite composed of a main body (mass m_0, centre of mass 0, principal axes of inertia x, y, z, principal moments of inertia A, B,

C) and of a small mass $m \ll m_0$ moving relatively to the satellite (Fig. 6.3). Elastic forces between m_0 and m do not exist. The influence of the stochastic motion of m on the motion of m_0 is considered .

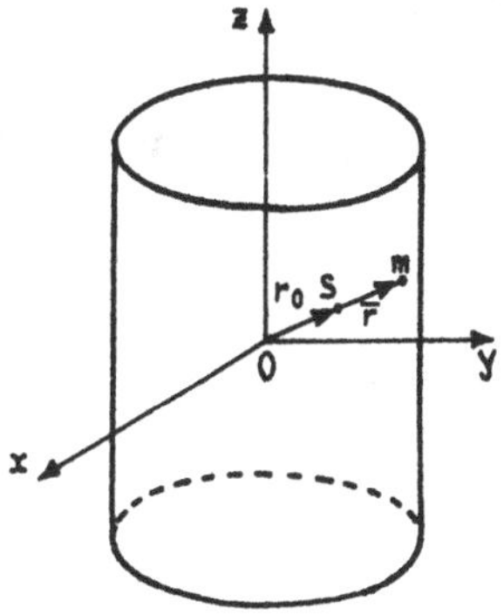

Fig. 6.3.

Satellite with stochastically moving additional mass

6. 2. 1 Moment of Momentum

The position of m relatively to 0 is described by the vector $\bar{r} = (r_x, r_y, r_z)$. The centre of mass of the whole system is S. The position of S is given by the vector

$$\bar{r}_0 = \frac{m}{m_0 + m}\bar{r} .$$

The absolute angular velocity of the main body may be $\bar{\omega} = (\omega_x, \omega_y, \omega_z)$. The moment of momentum of the system is

$$\bar{D} = \bar{D}_0 + \bar{D}_1 \tag{6.29}$$

where

$$\bar{D}_0 = \begin{bmatrix} A\omega_x \\ B\omega_y \\ C\omega_z \end{bmatrix} \tag{6.30}$$

and

$$\bar{D}_1 = m_0\bar{r}_0 \times \bar{v}_0 + m(\bar{r} - \bar{r}_0) \times (\bar{v} - \bar{v}_0) . \tag{6.31}$$

Here,

$$\begin{aligned} \bar{v}_0 &= \dot{\bar{r}}_0 + \bar{\omega} \times \bar{r}_0 \\ \bar{v} &= \dot{\bar{r}} + \bar{\omega} \times \bar{r} \end{aligned} \tag{6.32}$$

are the absolute velocities of S and m and the dot denotes the derivative in the x, y, z-system. It is

$$\begin{aligned} \bar{r}_0 &= \frac{m}{m_0 + m}\bar{r} \\ \bar{r} - \bar{r}_0 &= \frac{m_0}{m_0 + m}\bar{r} \\ \bar{v}_0 &= \frac{m}{m_0 + m}\bar{v} \\ \bar{v} - \bar{v}_0 &= \frac{m_0}{m_0 + m}\bar{v} \end{aligned} \tag{6.33}$$

and thus

$$\bar{D}_1 = m_0\frac{m}{m_0 + m}\bar{r}\times\frac{m}{m_0 + m}\bar{v} + \frac{mm_0}{m_0 + m}\bar{r}\times\frac{m_0}{m_0 + m}\bar{v} =$$

$$= \frac{m_0 m^2}{(m_0 + m)^2}\bar{r}\times\bar{v} + \frac{mm_0^2}{(m_0 + m)^2}\bar{r}\times\bar{v} = \frac{mm_0}{m_0 + m}\bar{r}\times\bar{v} =$$

$$= \frac{mm_0}{m_0 + m}\bar{r}\times(\dot{\bar{r}} + \bar{\omega}\times\bar{r}).$$

With the notation

$$\mu = \frac{mm_0}{m_0 + m} \tag{6.34}$$

we obtain finally

$$\bar{D}_1 = \mu\bar{r}\times(\dot{\bar{r}} + \bar{\omega}\times\bar{r}) \tag{6.35}$$

or

$$\bar{D}_1 = \mu\begin{bmatrix} r_y\dot{r}_z - \dot{r}_z\dot{r}_y + r^2\omega_x - r_x(\bar{\omega}\bar{r}) \\ r_z\dot{r}_x - r_x\dot{r}_z + r^2\omega_y - r_y(\bar{\omega}\bar{r}) \\ r_x\dot{r}_y - r_y\dot{r}_x + r^2\omega_z - r_z(\bar{\omega}\bar{r}) \end{bmatrix}. \tag{6.36}$$

6. 2. 2 Equation of Motion

Inserting (6.35) into the Euler equations we obtain

$$(6.37)\qquad \frac{d\bar{D}_0}{dt} + \bar{\omega}\times\bar{D}_0 + \frac{d\bar{D}_1}{dt} + \bar{\omega}\times\bar{D}_1 = \bar{M}_{ex}$$

where $\bar{M}_{ex}$ denotes the external torques. More complicated but in principle similar equations can be deduced for the case of several masses m_i moving relative to the main body [11].

6. 2. 3 Discussion of the Stochastic Models

The purpose is to describe the various inner motions of the satellite by means of one or more stochastically moving additional masses. In the case considered here the model is restricted to one mass m with the relative coordinates $r_x = r_x(t)$, $r_y = r_y(t)$ and $r_z = r_z(t)$. In different situations the assumptions on the stochastic processes $r_x = r_x(t)$, $r_y = r_y(t)$ and $r_z = r_z(t)$ will be different. For instance :

1. If only the existence of inner motions is known, but no further information is available, the whole expression $d\bar{D}_1/dt + \bar{\omega}\times\bar{D}_1$ can be taken as white noise. Thus, the effect of inner motions appears as a stochastic inner torque :

$$(6.38)\qquad \frac{d\bar{D}_0}{dt} + \bar{\omega}\times\bar{D}_0 = \bar{M}_{ex} + \bar{M}_i$$

with

$$\bar{M}_i = -\frac{d\bar{D}_1}{dt} - \bar{\omega}\times\bar{D}_1 = \dot{\bar{w}}\,.$$

2. Modeling the disturbances caused by the crew motions in the torus-shaped US 21 - man space station, it seems natural to assume m running randomly in a circle in the xy-plane (Fig. 6.4). In this case

$$r_x = r_0\cos\beta\,,\quad r_y = r_0\sin\beta\,,\quad r_z = 0\,,\quad r_0 = \text{const}\,.$$

and

$$\dot{r}_x = -r_0\dot{\beta}\sin\beta\,,\quad \dot{r}_y = r_0\dot{\beta}\cos\beta\,,\quad \dot{r}_z = 0\,.$$

Thus,

$$D_1 = \mu r_0^2\begin{bmatrix}\omega_x\sin^2\beta - \omega_y\sin\beta\cos\beta\\ \omega_y\cos^2\beta - \omega_x\sin\beta\cos\beta\\ \omega_z + \dot{\beta}\end{bmatrix}. \tag{6.39}$$

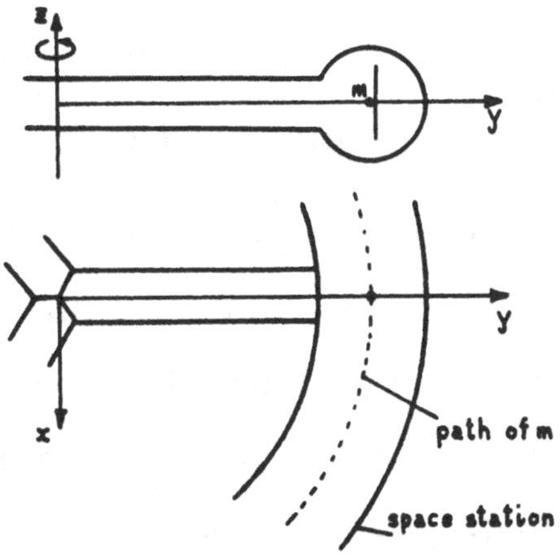

Fig. 6.4.

The 21-men space station.

Consider the pitch motion. The station is spin-stabilized and the nutations ω_x , ω_y can be supposed to be small. Then second order terms can be neglected. Regarding the symmetry of the station ($A = B$) the pitch motion is described by

$$\frac{dC\omega_z}{dt} + \mu r_0^2 \frac{d(\omega_z + \dot{\beta})}{dt} = 0 . \tag{6.40}$$

Integrating we obtain

$$\omega_z = \text{const.} - \frac{r_0^2 \mu}{C + r_0^2 \mu} \dot{\beta} . \tag{6.41}$$

Thus, the crew motion causes a fluctuation of the spin frequency.

Inserting (6.41) in the remaining two equations we get nonlinear stochastic equations for ω_x , ω_y :

$$\begin{aligned} d\omega_x &= f_1(\omega_x , \omega_y , \beta)dt + g_1(\omega_x , \omega_y , \beta)d\beta \\ d\omega_y &= f_2(\omega_x , \omega_y , \beta)dt + g_2(\omega_x , \omega_y , \beta)d\beta . \end{aligned} \tag{6.42}$$

The process $\beta = \beta(t)$ can be normally or Poisson distributed and must be fixed according to the information about the day schedule of the crew etc. For information on Wiener processes on manifolds (e.g. circles) see [25].

3. Suppose, m moves only in x-direction, i.e. $r_y = r_z = 0$. Then

$$D_1 = \mu r^2(t)\begin{bmatrix}0\\ \omega_x\\ \omega_y\end{bmatrix} \tag{6.43}$$

and the equations of motion are

$$\begin{aligned} A\dot{\omega}_x + (C - B)\omega_y\omega_z &= M_{xex}\,, \\ B\dot{\omega}_y + (A - C)\omega_z\omega_x &= M_{yex} + \mu r^2\omega_x\omega_z - \mu\frac{d}{dt}(r^2\omega_y)\,, \\ C\dot{\omega}_z + (B - A)\omega_x\omega_y &= M_{zex} - \mu r^2\omega_x\omega_y - \mu\frac{d}{dt}(r^2\omega_z) \end{aligned} \tag{6.44}$$

or

$$\begin{aligned} d\omega_x &= \left(-\frac{C - B}{A}\omega_y\omega_z + \frac{1}{A}M_{xex}\right)dt\,, \\ d\omega_y &= \left(-\frac{A - C}{B}\omega_z\omega_x + \frac{1}{B}M_{yex} + \frac{\mu}{B}r^2\omega_x\omega_z\right)dt - \frac{\mu}{B}d(r^2\omega_y) \\ d\omega_z &= \left(-\frac{B - A}{C}\omega_x\omega_y + \frac{1}{C}M_{zex} - \frac{\mu}{C}r^2\omega_x\omega_y\right)dt - \frac{\mu}{C}d(r^2\omega_z)\,. \end{aligned} \tag{6.45}$$

This system has to be completed by the definition of the process $r^2 = r^2(t)$, for instance by a shaping filter

$$dr^2 = f\,dt + g\,dw\,. \tag{6.46}$$

By comparison of the three examples considered above, the possibilities and bounds of the theory of stochastic stability can be demonstrated. In the first case we obtained a usual stochastic differential equation (6.38). The corresponding linearized equation can be solved using formula (2.44). The stability of the nonlinear eq. (6.38) can be investigated by Liapunov techniques. In the case of the 21-man space station, the coefficients of the resulting equations (6.42) depend on β and thus eqs. (6.42) are not of the standard form (2.36). However, the reduction to the standard form can be achieved augmenting the state vector (ω_x , ω_y) by the noise β to the vector (ω_x , ω_y , β). In consequence, the stability check is considerably complicated. In the third case the differentials $d(r^2\omega_x)$, $d(r^2\omega_y)$ occur in the resulting eqs. (6.45). In general, the reduction of the unusual system (6.45), (6.46) to the standard form (2.36) is not possible. In place of eq. (2.36) more general forms of stochastic equations must be taken into consideration. Some of these forms are treated in [4].

References

[1] Itô, K. : On Stochastic Differential Equations, Memoirs of Amer. Math. Soc., 1951.

[2] Bucy, R.S., Joseph, P.D. : Filtering for Stochastic Processes with Applications to Guidance, J. Wiley & Sons, 1968.

[3] Kushner, H.J. : Stochastic Stability and Control, Academic Press, 1967.

[4] Gikhman, J.J., Skorokhod, A.V. : Stochastic Differential Equations (Russ.) Kiev, 1968.

[5] Khas'minski, R.S. : The stability of Systems of Differential Equations With Random Disturbances of Parameters (Russ.),Moscow, 1969.

[6] McKean, H.P. : Stochastic Integrals, Academic Press, 1969.

[7] Stratonovich, R.L. : A New Representation for Stochastic Integrals and Equations, Siam J. Control 4, 1966.

[8] Loève, M. : Probability Theory, Van Nostrand Co., 1963.

[9] Doob, J. : Stochastic Processes, J. Wiley & Sons, 1960.

[10] Dynkin, E.B. : Die Grundlagen der Theorie der Markoffschen Prozesse, Springer, 1961.

[11] Aleksejev , K.B., Bebenin G.G. : Space Vehicle Control, NASA TTS-9336, Ohio, 1966.

[12] Beletzkij, V.V. : Motion of an Artificial Satellite About its Centre of Mass, Jerusalem, 1966.

[13] Sagirow,P.S., Satellitendynamik, BI-Verlag, 1970.

[14] Schrello, D.M. : Dynamic Stability of Aerodynamically Responsive Satellites, Journ. Aerosp. Sci., 1962.

[15] Sheporaitis, L.P. : Stochastic Stability of a Satellite Influenced by Aerodynamic and Gravity Gradient Torques,AIAA 8th Aerosp. Sci. Meeting, New-York, Jan. 1970.

[16] Hahn, W. : Stability of Motion, Springer, 1967.

[17] Kalman, R.E., Falb, P.L., Arbib, M.A. : Topics in Mathematical System Theory, McGraw-Hill, 1969.

[18] Loh, W.H.T. : Dynamics and Thermodynamics of Planetary Re-entry, Prentice-Hall, Englewood Cliffs., 1963.

[19] Citron, S.I., and Meir, T.C. : An Analytic Solution for Entry into Planetary Atmospheres, AIAA Journ., Vol. 3, No. 3, 1965.

[20] Handbook of Geophysics and Space Environments, McGraw-Hill, 1966.

[21] King-Hele, D. : Theory of Satellite Orbits in an Atmosphere, Butterworths, 1964.

[22] Weiss, G.R. : Re-Entry Dispersions Due to Atmospheric Uncertainties, Journ. of Spacecraft and Rockets, Vol. 6, Nr. 10, 1969.

[23] Krementulo, V.V. : On Optimal Stabilization of a Rigid Body with Fixed Point by Means of Flywheels, PMM 30, 1966.

[24] Pestel, E.D., Leckie, F.A. : Matrix Methods in Elastomechanics, McGraw-Hill, 1963.

[25] Itô, K. and McKean, H.P. : Diffusion Processes and Their Sample Paths, Springer, 1965.

Contents

	Page
Preface	3
Chapter 1. Introduction	5
Chapter 2. Stochastic Dynamical Systems	8
2.1. Basic Definitions	9
2.1.1. Probability Density Functions	9
2.1.2. Moments	10
2.1.3. Random Vectors	10
2.1.4. Stochastic Processes	12
2.1.5. Markov Processes	13
2.1.6. Processes with Independent Increments	13
2.1.7. Wiener Processes	14
2.1.8. White Noise	14
2.1.9. Diffusion Processes	16
2.2. Stochastic Dynamical Systems	16
2.3. Stochastic Integrals	19
2.4. Stochastic Differentials	21
2.5. Itô Formulae	21
2.5.1. dg(w)	21
2.5.2. dg(t,x(t))	23
2.5.3. dx_1x_2	24

2.6. Stochastic Differential Equations.... 25

2.6.1. Definitions and Notations........ 25

2.6.2. General Properties................ 27

2.6.3. Explicit Solutions of Linear Equations.......................... 28

2.7. Stochastic Stability................. 30

2.7.1. Stability in the Mean............ 31

2.7.2. Stability in the Mean Square..... 32

2.7.3. Two Theorems on the Stability of the Moments........................ 36

2.7.4. Definitions of Stochastic Stability.............................. 37

2.7.5. Intuitive Approach to the Stochastic Liapunov Method.......... 39

2.7.6. Strong Formulation of the Stochastic Liapunov Method.......... 41

2.7.7. The Operator L.................... 42

2.7.8. An Important Special Case........ 43

2.7.9. Attraction Domains................ 44

2.7.10. Stability Check via Linearized Equation.......................... 45

Chapter 3. Orbiting Satellite Influenced by Gravity Gradient and Aerodynamic Torques.............................. 48

3.1. Librational Motion.................... 48

3.2. Gravity Gradient Torques............. 50

3.3. Aerodynamic Torques.................. 51

3.4. Stochastic Differential Equation of the Librational Motion.............. 52

3.5. Stability of the Pitch Motion........ 58

3.5.1. Stability Conditions via Covariance Equation........................ 58

3.5.2. Stability Conditions via Liapunov Techniques....................... 67

3.5.3. Discussion....................... 74

Chapter 4. Dispersions of the State Variables During Re-Entry....................... 77

4.1. Re-Entry Equations................. 78

4.2. The Stochastic Model of the Atmospheric Density....................... 80

4.3. Fluctuations Caused by Density..... 83

4.4. Simplified Models.................. 85

4.5. Covariance Equation................ 89

4.6. Dispersions of Arc Length and Velocity........................... 90

Chapter 5. Stabilization by the Magnetic Field of the Earth......................... 92

5.1. Satellite in Equatorial Orbit...... 92

5.2. Fluctuations of the Magnetic Field of the Earth [20] and the Stochastic Model of Yaw Motion............... 94

5.3. Stability of the Yaw Motion........ 95

5.3.1. Stability of the Second Moments 96

5.3.2. Stability of the Nonlinear Equation.......................... 96

5.3.3. Discussion.................... 99

Chapter 6. Satellites With Moving Parts......... 101

6.1. Stabilization by Wheels With Noisy Frictional Moment.................. 101

6.1.1. The Deterministic Model....... 102

6.1.2. The Stochastic Model.......... 108

6.1.3. Stability and Errors.......... 111

6.2. Satellite with Variable Moments of Inertia............................ 116

6.2.1. Moment of Momentum............ 117

6.2.2. Equation of Motion............ 119

6.2.3. Discussion of the Stochastic Models........................ 120

References................................. 125

Printed in the United States
By Bookmasters